TUNA ISSUES AND PERSPECTIVES IN THE PACIFIC ISLANDS REGION

Edited by
David J. Doulman

David J. Doulman is a fellow and the director of the Pacific Islands Development Program's tuna project at the East-West Center, Honolulu. Before joining the program in 1985, he was chief fisheries economist for the Papua New Guinea government. Doulman received his doctorate from James Cook University, Australia, and has written widely on tuna and fisheries in the Pacific islands region.

Library of Congress Cataloging-in-Publication Data

Tuna issues and perspectives in the Pacific Islands
  region.

  1. Tuna fisheries—Islands of the Pacific. I. Doulman,
David J., 1950–
SH351.T8T785 1987   338.3'72758   87-5372
ISBN 0-86638-093-0

CONTENTS

Foreword
I. T. Tabai   v

Acknowledgments   vii

Editor's Introduction   ix

PART I. INTRODUCTION

1. Tuna Fisheries Management in the Pacific Islands Region
   Parzival Copes   3

2. Tuna and the Impact of the Law of the Sea
   Anthony J. Slatyer   27

3. Biological Perspectives on Future Development
   of Industrial Tuna Fisheries
   John Sibert   39

PART II. DISTANT-WATER TUNA FISHERIES

4. Development of Japan's Tuna Fisheries
   Norio Fujinami   57

5. Postwar Development and Expansion
   of Japan's Tuna Fishery
   Yoshiaki Matsuda   71

6. U.S. Tuna Fleet Ventures in the Pacific Islands
   August Felando   93

7. U.S. Tuna Policy: A Reluctant Acceptance
   of the International Norm
   Jon M. Van Dyke and Carolyn Nicol   105

8. Development and Expansion of the Tuna
   Purse Seine Fishery
   David J. Doulman   133

9. Distant-Water Tuna Longline Fishery
   Michael J. Riepen   161

PART III. ARTISANAL AND DOMESTIC TUNA FISHERIES

10. The Importance of Small-Scale Tuna Fishing:
A Tokelau Case Study
Robert Gillett and Foua Toloa   177

11. American Samoa: The Tuna Industry and the Economy
Donald M. Schug and Alfonso P. Galea'i   191

12. High Speed on an Unmade Road:
Solomon Islands' Joint-Venture Route to a Tuna Fishery
Anthony V. Hughes   203

13. Growth and Contraction of Domestic Fisheries:
Hawaii's Tuna Industry in the 1980s
Linda Lucas Hudgins and Samuel G. Pooley   225

PART IV. REGIONAL AND INTERNATIONAL ASPECTS

14. History and Role of the Forum Fisheries Agency
Florian Gubon   245

15. Fisheries Cooperation: The Case of the Nauru Group
David J. Doulman   257

16. Global Tuna Markets: A Pacific Island Perspective
Dennis M. King   279

PART V. FUTURE DIRECTIONS

17. Prospects and Directions in the Tuna Fishery
David J. Doulman   299

Contributors   313

Foreword

It is a great honor to have been asked to write a foreword to this publication on tuna and tuna-related issues. Fishing has always been—and still is—vital to the subsistence way of life of South Pacific peoples. Recently, however, they have recognized that fishing—tuna fishing in particular—holds the greatest potential for economic development and economic self-reliance to support their recently acquired political independence.

The economic significance of tuna as a resource to the island nations did not come to be fully recognized until the United Nations Convention of the Law of the Sea was adopted in early 1982. This Convention enabled countries bordering the sea to extend their territorial sea boundaries to 12 nautical miles and their exclusive economic zones (EEZs) to 200 nautical miles. The result is that South Pacific countries now control millions of square kilometers of ocean, including the tuna resources that have long been exploited mainly by the developed distant-water fishing nations, who have paid little or nothing to exploit them.

This new regime introduces factors that strain relations between the Forum Island Countries (FICs) and the distant-water fishing nations because outside fishermen naturally want to pay as little as possible, whereas the FICs want the very opposite. This potential for strained relations became evident at the time the Forum Fisheries Agency (FFA) was created, when it was finally decided that the agency's sole purpose would be to assist the forum countries to get the maximum benefit from their tuna resources. The agency's policies and their implementation are discussed from differing perspectives in this book.

However, it is one thing for the island nations to set a policy; it is another to achieve it. And though there is no denying that benefits have begun to accrue, it cannot yet be claimed that a fair return has been achieved. Indeed, there is still a long way to go. The situation is unlikely to change much in the near future because it requires negotiation, and nations in our region have little bargaining power relative to the major fishing nations, from which they receive substantial assistance in financing and promoting various regional and national development programs, including fisheries development. It is these same countries that have the technology to exploit the tuna resources of the Pacific region.

Despite the problems facing the island countries, there is one thing certain: the South Pacific countries are committed to working to achieve what they believe to be a fair return from their tuna resource. This commitment is the reason for the creation of the FFA and the Nauru Group, both of which have assisted the forum countries in cooperating more effectively against the bargaining power of the distant-water fishing nations. More recently these countries have shown their willingness to stand up for their own national interests, even at the cost of external opposition and strained relations with countries traditionally regarded as their friends. They have taken the first independent step that can make the next steps easier.

To the Pacific countries, the whole question of tuna is about economics, not ideology. They want—and are working for—economic security for their people. It is for this reason that they are developing locally based fishing industries and are continuing to struggle in the face of harsh economic conditions in the tuna industry. The experiences of some of these countries are related in this book. The point to note about the Pacific island states is their keen desire for mutually beneficial relationships with other countries, particularly the traditional friends with whom they have much in common. The relationship must, of course, be a balanced one, fair not only in appearance but in actual fact.

This book is the first on this subject ever to be published. It attempts to present a wide range of views on the many aspects of tuna fishing, with emphasis on the South Pacific region. It is appropriate to address this subject now, because tuna fishing is and will be important, not only for the economic development of the region but also for its implications for relations between the South Pacific countries and the distant-water fishing nations.

More important, however, the book will contribute to a wider understanding of the many perspectives on issues related to tuna fishing. Such understanding can enhance cooperative efforts in the orderly exploitation and management of the tuna resource for the benefit of our region and foster the future well-being of the Pacific. The authors are to be congratulated for their effort. I commend this publication to all who have interest in tuna fishing and in the economic well-being of Pacific peoples.

I. T. Tabai
Beretitenti
Republic of Kiribati
August 1985

Acknowledgments

This book brings together for the first time a collection of papers focusing on socioeconomic issues related to tuna fisheries in the Pacific islands region. A group of diverse professionals have contributed to the book, providing perspectives from industry, government, and the academy. Of course, not all contributors agree; their philosophies differ depending on whether they represent the interests of island countries, distant-water fishing nations, industry, government, or regional organizations. However, the compilation of a range of views and opinions under one cover is a healthy indication that various actors in the industry are both able and willing to engage in constructive dialog—dialog that can only serve the interests of all those involved in exploiting and managing the Pacific islands' tuna resources.

All contributors to the book participated enthusiastically, without remuneration, and in spite of heavy work commitments. I am indebted to them because this book would not have been possible without their support. Production of the book took longer than expected for a variety of reasons, not the least of them the vast geographic distribution of contributors.

I am also indebted to Charles W. Lepani, director of the East-West Center's Pacific Islands Development Program; Jeanne Hamasaki, former program officer; and Michael Hamnett, former research coordinator of the program—all of whom encouraged and facilitated the production of the book. Jane Aucoin, Titilia Barbour, Forrest Hooper, Lynette Tong, and Mary Yamashiro gave secretarial support. Edith Kleinjans edited the contributions, and Jacqueline D'Orazio and her staff managed production.

Finally, without the support of the USAID regional director, Bill Paupe, publication of the book would have been difficult. Its publication was generously supported by Grant No. 879-0010-G-SS-6021-00 from the South Pacific Regional Development Office, United States Agency for International Development, Suva, Fiji.

David J. Doulman
Honolulu

March 1987

Editor's Introduction

The Pacific islands region is a large and diverse area of the western and central Pacific Ocean stretching from the Republic of Palau in the west to Pitcairn Island in the east. In this region are 14 independent or self-governing countries and 8 territories of France, New Zealand, the United Kingdom, and the United States. Except for Papua New Guinea, Solomon Islands, and Fiji, each of the countries and territories consists of a single small island or a group of sparsely distributed islands (Figure 1).

All these countries and territories are marked by high economic dependence. The situation of most of them is likely to worsen because (1) they have narrow economic bases and (2) their agricultural resources and land-based economic activities are limited. For these reasons their

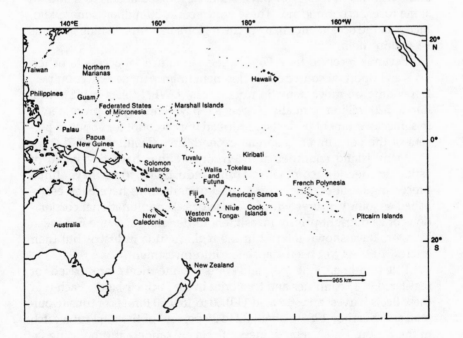

Figure 1. Pacific islands region

marine resources are central to the subsistence and culture of their peoples.

In Micronesian and Polynesian countries and territories a paucity of inshore fisheries has traditionally required fishermen to exploit pelagic resources, especially tuna, for their livelihood. This reliance continues. However, subsistence exploitation of tuna is small relative to the size of the resource and its exploitation by commercial fishing fleets. Doulman and Kearney (1986) estimate that subsistence catches of tuna in the region do not exceed 10,000 tonnes annually.

Tuna is the most abundant and most valuable commercial fisheries resource in the region. Although precise data on commercial catches are not available, estimates are that 650,000 tonnes of tuna with an ex-vessel market value of at least $700 million was harvested there by domestic fleets and the fleets of distant-water fishing nations (DWFNs) in 1984 (Doulman 1987a).

Japanese fishermen started exploiting the tuna resources of the region in the early 1900s. Since then commercial fisheries have developed and expanded. The world's leading DWFNs—Japan, Korea, Taiwan, and the United States—operate throughout the region. Their fleets fish in the exclusive economic zones (EEZs) of island countries and territories in return for payment of negotiated access fees and, in some cases, economic aid. These zones cover 30 million square kilometers of the ocean, and many of the region's best tuna fishing grounds lie within them.

Revenue received from DWFN fleets in return for access to fisheries is an important source of development finance for several countries. Depending on movements in tuna prices, DWFN fleets pay $15 million to $20 million annually. Access fee payments figure prominently in some government budgets; in Kiribati they account for about 25 percent of the government's annual expenditures (Doulman 1987b).

Many island countries and territories either have developed or aspire to develop commercial tuna industries; some have tuna-processing facilities. Such commercial undertakings have been established as joint ventures with Japanese or U.S. multinational corporations or as outright private investment undertakings. Some European investors have shown interest in the region's tuna industry, but their interest has not yet been translated into investment.

Pole-and-line, longline, and purse seine fleets are based or registered in 13 countries and territories in the Pacific islands. Each year these fleets harvest an estimated 140,000 to 160,000 tonnes of tuna (Doulman and Kearney 1986)—about 20 to 25 percent of the total tuna catch in the region. Catches by domestic fleets are landed at processing facilities in the region or consolidated at transshipment locations for export to processors outside the region. At present, tuna is transshipped at Guam, Tinian (Northern Marianas), Rabaul (Papua New Guinea), Tulagi (Solomon Islands), and Palikula (Vanuatu). Before 1982 Palau also served as a transshipment base. Several other island countries and territories are investigating the possibility of establishing shore-based transshipment facilities, primarily to service DWFN fleets operating in their waters.

Tuna landed in the Pacific islands is either canned or processed into *katsuobushi* (smoke-dried skipjack). *Katsuobushi* is processed in the Marshall Islands and Solomon Islands, then exported to Japan, where it is just a small proportion of the total market. *Katsuobushi* production is constrained by the size of the Japanese market (about 32,000 tonnes a year) and by the shortage of suitable timber for smoking the product, particularly in countries consisting of atolls.

There are tuna-canning facilities in American Samoa, Fiji, and Solomon Islands. The two canneries in American Samoa are among the world's largest, with a combined annual processing capacity of 155,000 tonnes. These canneries account for nearly 90 percent of the region's canning capacity. The canneries in Fiji and Solomon Islands can process 15,000 tonnes and 5,000 tonnes a year respectively. Hawaii's cannery, which operated until 1984, had an annual processing capacity of 35,000 tonnes. This cannery is scheduled to reopen in 1987, but on a significantly reduced scale. Solomon Islands is relocating and expanding its cannery. Several other island countries are planning to develop tuna-processing facilities.

The economies of three Pacific island countries depend heavily on their tuna industries (Doulman and Kearney 1986). Tuna processing is American Samoa's dominant industry. Processed tuna now accounts for about 90 percent of the value of American Samoa's exports, 30 to 40 percent of Solomons Islands' exports, and about 30 percent of Vanuatu's exports.

The politically independent countries in the region cooperate closely on fisheries matters through the South Pacific Forum Fisheries Agency (FFA). Since 1979 this cooperation has permitted them to present a united front in dealing with DWFNs and to secure a fairer share of the benefits flowing from the exploitation of their tuna resources. Hawaii and the U.S. territories in the Pacific also cooperate on fisheries matters through the Western Pacific Regional Fishery Management Council. However, the U.S. position on tuna has impeded the council's efforts to develop a management plan for tuna.

The extent of dependence on tuna in the Pacific islands is unmatched elsewhere. Pacific islanders depend on tuna for subsistence; island governments depend on revenue from DWFNs to support their development programs; the economies of several countries and territories depend on tuna fishing, tuna processing, and associated industries. Because of this dependency on tuna and the revenue it generates—and the lack of alternative development avenues—the ownership of tuna and the right to harvest it are sensitive political issues in the Pacific islands. The past failure of the United States to recognize the rights of Pacific island countries to tuna ownership alienated them and eroded their good will toward the United States. Economically and financially, this failure was not in the interests of Pacific island countries; politically, it worked against broader U.S. interests in the region. It has also fueled superpower rivalry in the region, adding yet another dimension to a complex situation.

Such are the circumstances surrounding the tuna industry in the Pacific islands region, which is the substance of this book. The book discusses the development of the industry from a variety of perspectives, explores sundry issues arising within it, and speculates on its prospects.

This book consists of 17 chapters arranged in five sections. The first section presents management concepts, legal aspects of the Law of the Sea, and biological perspectives. This section provides a backdrop for subsequent sections. In it Parzival Copes discusses fisheries management and describes how management concepts might be applied to the tuna fishery in the Pacific islands region. Anthony Slatyer analyzes provisions of the Law of the Sea and their ramifications for development and management, and John Sibert discusses biological issues pertinent to the development of industrial tuna fishing in the islands region.

The second section, with six chapters, examines aspects of distant-water tuna fisheries in the Pacific islands region. Norio Fujinami reviews the development of Japan's tuna fisheries, focusing on government policy and institutional arrangements for promoting and managing the country's tuna industry. Yoshiaki Matsuda's chapter traces the expansion of Japan's tuna fisheries after World War II and discusses factors that shaped Japan's tuna industry. The next chapter, by August Felando, outlines the contribution of U.S. fishermen to the development of the tuna fishery in the Pacific islands. This chapter is complemented by Jon Van Dyke and Carolyn Nicol's analysis of U.S. tuna policy, the problems it caused in the islands region, and the evolution of the U.S. multilateral tuna treaty. David Doulman then reviews the development and expansion of the purse seine tuna fishery in the Pacific islands, and Michael Riepen examines the development and current status of the region's distant-water longline fishery.

The third section has four chapters exploring issues in artisanal and domestic tuna fisheries. Robert Gillett and Foua Toloa present a case study of small-scale tuna fishing in Tokelau. The chapter by Donald

Schug and Alfonso Galea'i analyzes the economic importance of American Samoa's tuna industry. Anthony Hughes then evaluates Solomon Islands' contrasting experience with the development of its tuna industry. Finally, Hawaii's tuna industry, now in a state of flux, is reviewed by Linda Lucas Hudgins and Samuel Pooley.

Regional and international aspects of the tuna industry in the Pacific islands are covered in three chapters in the fourth section. Florian Gubon presents the history of the Forum Fisheries Agency and its role in the region. Regional cooperation in the Nauru Group is analyzed and evaluated in David Doulman's chapter. Dennis King then reviews world tuna markets from the perspective of Pacific island countries.

The final section is about future directions in the tuna fishery. David Doulman speculates on a range of issues likely to affect developments in the tuna fishery, including the position of DWFN and domestic fleets, processing industries, and regional cooperation.

To provide a standard for comparison, all values are given in U.S. dollars and all measures in metric units.

REFERENCES

Doulman, David J.

1986   Fishing for tuna: The operation of distant-water fleets in the Pacific islands region. Research report series no. 3. Pacific Islands Development Program. East-West Center. Honolulu. 38 pp.

1987a   The tuna industry in the Pacific islands region: Opportunities for foreign investment. Marine Fisheries Review. In press.

1987b   The Kiribati-Soviet Union fishing agreement. Pacific Viewpoint. In press.

Doulman, David J., and Robert E. Kearney

1986   The domestic tuna industry in the Pacific islands region. Research report series no. 7. Pacific Islands Development Program. East-West Center. Honolulu. In press.

PART I. INTRODUCTION

1.
Tuna Fisheries Management in the Pacific Islands Region

Parzival Copes

INTRODUCTION

In fisheries management the textbook case is a single-species fishery on a stock confined to the waters of a single country, conducted by fishing units of a single gear type fully dedicated to that fishery.[1] The tuna fishery of the Pacific islands region meets none of these criteria of simplicity. The vessels in this fishery come from many countries, near and far. They pursue two major and several minor species of tuna across the maritime zones of the region's 22 states and other political divisions—a vast area of over 30 million square kilometers (Doulman 1986a). Moreover, these tuna stocks migrate far beyond the region across wide stretches of the high seas and the maritime zones of many countries. Vessels of at least three distinct gear types participate in the fishery. The circumstances under which the fishery for tuna is conducted in the region are complex, making effective management difficult to achieve.

Managing a highly migratory stock such as tuna requires consistent control over the fishery on that stock throughout its whole migration range. It calls for the cooperation of all the resource-adjacent nations (RANs) through whose maritime zones the fish migrate. Where the stock also passes through the international waters of the high seas, the cooperation of distant-water fishing nations (DWFNs) whose fleets fish there is also needed.

The RANs of the Pacific islands region include 14 small independent or self-governing states and 8 territories (Doulman 1986a). The former include the Cook Islands, the Federated States of Micronesia, Fiji, Kiribati, the Marshall Islands, Nauru, Niue, Palau, Papua New Guinea, Solomon Islands, Tonga, Tuvalu, Vanuatu, and Western Samoa.[2] Most of them are members of a regional organization, the South Pacific Forum (SPF); the others have observer status. The territories include three dependencies of France (French Polynesia, New Caledonia, and Wallis and Futuna), three of the United States (American Samoa, Guam, and the Northern Marianas), and one each of Great Britain (Pitcairn Island) and New Zealand (Tokelau).

There are locally based tuna fleets in 11 of the 22 states and territories of this region (Doulman 1985). Their annual catch is in the range of 80,000 to 100,000 tonnes out of a total regional catch estimated at 630,000 tonnes in 1984. Thus the bulk of the catch is taken by the fleets of DWFNs. Japan and the United States take by far the biggest catches, but South Korea and Taiwan also take large harvests. Much smaller amounts are taken by a number of other countries, sometimes only on an occasional basis. The Forum Fisheries Agency (FFA), acting on behalf of SPF members, maintains a register of vessels permitted to fish in its members' zones. Apart from the four major DWFNs, the countries with registered vessels include the Cayman Islands, Honduras, Indonesia, Mexico, New Zealand, Panama, the Philippines, and the Soviet Union (Doulman 1986a, 8).

The tuna stocks of the Pacific islands region migrate into adjacent areas of the high seas, where DWFNs may fish without authorization. The number of additional DWFNs that might participate in the fishery on tuna stocks of the area could therefore be substantial, compounding the complexity and difficulty of managing the tuna stocks in the Pacific islands.

This paper explores in a general fashion the purposes, problems, possibilities, and prospects of managing international tuna fisheries in the Pacific islands region. After a discussion of the goals of management, a theoretical overview of the bioeconomic principles underlying fisheries management is presented, with emphasis on points of particular relevance to tuna management in the region. To accommodate a diverse readership, the theoretical presentation is kept as simple as possible. (Those interested in a more extensive analysis and fuller explanation may consult the references listed.) Later sections consider the implications of new concepts and rules in the Law of the Sea for international tuna management in the context of the Pacific islands region and analyze the practicalities of available management options and the international arrangements necessary for success.

GOALS OF FISHERIES MANAGEMENT

There is now broad recognition of the need for special arrangements for managing fish stocks that are not appropriate for most other resources, simply because most fish stocks are common property whereas most other resources are privately owned or controlled. A farm or forest is managed by the owner or leaseholder to maintain its productivity and yield a good economic return because of the owner's or leaseholder's stake in it. But a fisherman who neither owns nor controls the stock he fishes has no incentive to maintain its productivity by limiting his harvest or investing in the stock's enhancement, because he cannot capture the benefits of such actions. Rival fishermen with equal access to the enhanced stock would simply increase their catches and reap most of the rewards. Because fish stocks are common property, they are easily exploited by harvesters competing for catches without regard for the impact on stock regeneration. In the long run this kind of exploitation is bound to depress the productivity of the stock and the economic returns from the catch.[3]

The pernicious results of unregulated fishing have increasingly led governments to act as surrogate owners by assuming responsibility for managing the resource. It is widely recognized that if the problem of common-property fisheries is overexploitation, the solution must be sought in limiting the number of fishing units and the effort they put out. Such a limitation of effort offers a two-sided advantage: it cuts the aggregate costs of fishing, and it reduces pressure on fish stocks so that they can be restored to yield larger catches and revenues. In the analytical terms of economics, limiting effort to an optimum level allows maximum net economic benefits to be extracted from the fishery (Copes 1972a). In particular, it allows for the regeneration of resource rents that are dissipated by the excessive fishing effort of an open-access fishery (Gordon 1954). The resource rents generated may be left with fishing enterprises to enhance their incomes; captured as taxes, fees, or royalties by governments to add to their public revenue; or shared by government and industry.

Over the past few decades several countries have developed sophisticated management regimes for domestic fish stocks, often achieving effective stock conservation. In many cases economic returns have also improved substantially. Still, much remains to be done in refining management techniques and overcoming political and social obstacles to obtain optimum results.

BIOLOGICALLY ORIENTED MANAGEMENT CONTROLS

Over the years a variety of fisheries management techniques have been put into practice. Initially, governments were concerned primarily with stock conservation; hence they drew their advice primarily from biologists. Conservation of a fish stock requires that losses from natural mortality and fishing mortality be counterbalanced by growth in the stock. For management purposes, a fish stock is usually considered to consist only of individual fish that have reached "fishable" size. Following that criterion, one may distinguish between two sources of growth in a stock: (1) recruitment, the process by which juvenile fish reach fishable size and become part of the stock, and (2) the gain in biomass resulting from the growth of fish already part of the stock.

A number of regulations have been developed to promote adequate stock growth. Some are designed specifically to support recruitment. One such measure is closing an area of a fishery for a defined period of time to protect spawning concentrations, to allow migrating spawners to escape, and to safeguard nursery areas. Other measures are designed to increase individual growth by methods that support selectivity in harvesting. Younger and smaller fish that have high growth rates are allowed to escape so that they may be available to the fishery at a larger size. The optimal minimum size at first capture is associated with a high (or maximum) yield per recruit. The selective escapement of smaller fish may be promoted by prescribing minimum mesh sizes for net fishing, minimum hook sizes for line fishing, and minimum escape gaps for trap fishing.

Some management measures may assist both recruitment and individual growth simply by containing fishing effort to lessen pressure on the stock. In many fisheries a total allowable catch (TAC) is set each season (usually a year). When the TAC is reached, the fishery is closed for the remainder of the season.

For analyzing the effects of fishing effort on fish catches over time, biologists have developed explanatory models.

The most widely applied model originates with Schaefer (1954), who used the assumption of a logistic growth function to derive the parabolic yield-effort relationship portrayed in (A) in Figure 1. The model shows the yields that can be sustained (per season/year) at various levels of long-run effort.[4] This long-run qualification is important. The short-run effect of any increase in effort, of course, is to increase the catch. But a concomitant effect of the increased catch is a decrease in the standing stock of fish, leading to lower catches in subsequent years. It is only after the same level of effort has been applied for a number of years (the number depending on the stock dynamics) that a new equilibrium is established among effort, stock, and catch levels. In the long run, high levels of effort result in low catches because of the depletion of the fish stock.

The Schaefer model predicts extinction of a stock (zero yield) at a finite level of effort (*OX* in (A) in Figure 1) and a maximum sustainable yield (MSY) at a level of effort (*OK*) that is half of the effort level at the point of stock extinction (*OX*). Several studies and observations have suggested that most large and widely distributed fish stocks cannot in practice be fished to extinction—certainly not at a level of effort only

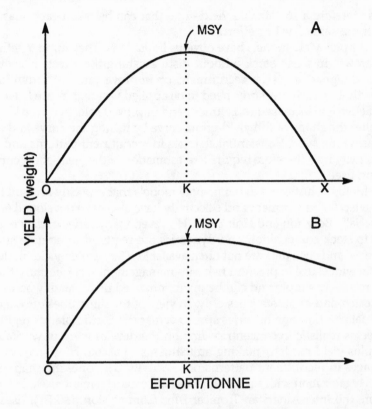

Figure 1. Yield-effort relationships derived from a logistical growth function (Panel A) and a Gompertz growth function (Panel B)

twice that of the MSY. This finding calls into question the simple symmetrical form of the Schaefer yield curve. Fox (1970) has suggested an alternative model with an open-ended yield curve based on a Gompertz growth function. His model is illustrated in (B) in Figure 1. The Fox model is probably more appropriate than the Schaefer model for most fish stocks, certainly for the widely distributed tuna stocks, which show declining yields but no sign of approaching extinction at high levels of effort.

The Schaefer and Fox curves are both surplus-yield models. The sustainable yields measured by these curves represent the net growth (total growth minus mortality) per time period in the fish stock at different population levels during different levels of long-run effort. The net growth in the stock at any particular population level, then, represents a surplus that can be harvested without reducing the population level.

It is therefore a sustainable yield—one that can be taken each season with the same level of effort.

Surplus-yield models have obvious limitations. They lump together all growth in a fish stock without distinguishing the effects of measures designed to assist recruitment or improve rates of individual growth. But such measures need to be applied where they are feasible and effective. Increased recruitment and improved selectivity are likely to alter the shape of the yield-effort curve by raising the curve and increasing the MSY. Substantial changes in recruitment patterns and in selectivity may therefore require a re-estimation of the yield-effort curve being used.

More sophisticated dynamic-pool models that explicitly account for the effects of recruitment and selectivity have also been developed (see especially Beverton and Holt 1957). However, many series of data relating to stock composition and dynamics are required to operate such models, and such data are not often available. The surplus-yield models are in widest use in practical fisheries management work because they are relatively simple and can be operationalized using readily accessible catch-and-effort statistics only. In view of the difficulties of collecting data on the tuna fisheries spread across the Pacific islands region, it seems realistic to concentrate first on the development of workable surplus-yield models, making adjustments to account for significant changes in recruitment patterns or selectivity. The only international tuna organization with a substantial in-house management research program, the Inter-American Tropical Tuna Commission (IATTC), uses a surplus-yield model for its basic analytical overview.

Both the Schaefer and Fox models are deterministic; that is, they predict catches from given long-run effort levels. But because these models cannot distinguish the effects of changing recruitment and selectivity patterns, they may fail to predict catches accurately. The innumerable environmental factors that affect fish stocks, many of them random and unpredictable, further reduce the accuracy of these models' predictions. The usefulness of the models for a particular stock thus depends very much on the stock's sensitivity to environmental variations. Some sensitive stocks show enormous variations in recruitment from year to year, and the stock strength of short-lived species can vary over shorter periods of time, regardless of fishing activity. Surplus-yield models are next to useless in managing such stocks. Fortunately, tuna stocks are long-lived and sufficiently robust to withstand most random environmental influences. The surplus-yield models therefore are reasonably well suited to use in tuna fisheries management.

An obvious feature of the surplus-yield model is its clear identification of the maximum sustainable yield (MSY). Regulation of the fishery so as to achieve a harvest level equal to the MSY has often been put forward as the major management objective in both national and international fisheries because the MSY measures the largest amount of fish that can be taken without risk of stock depletion. In a world with many hungry people suffering protein deficiency, extracting a maximum yield from the fishery on a sustained basis has both moral and rational appeal. Moreover, the MSY goal appears achievable by relatively simple regulatory means. Effort may be regulated directly to be consistent with MSY, or it may be constrained indirectly by enforcing a TAC equal to MSY. If stocks have been overfished, they will have to be built up by temporarily reducing TACs or placing more severe constraints on effort.

BIOECONOMIC MANAGEMENT

Economists began showing an analytical interest in fisheries in the mid-1950s and became involved in fisheries management during the 1960s. Recognizing the importance of stock behavior under exploitation, economists incorporated basic elements of fish population dynamics in their models, thereby creating bioeconomic models.

From the beginning, economists have questioned the validity of MSY as a goal of fisheries management, pointing out that it indicates only gross benefits from the fishery, not what it is worth or what it costs to produce. From an economic standpoint, the true measure of benefits from the fishery is a net measure showing the excess of the value of the catch over its cost of production. This is illustrated in simplified form in Figure 2.[5] If it is assumed that any output from a fishery can be sold at the same price, the yield curve can be transformed into a revenue curve by multiplying the yield at different effort levels by the constant price. The same curve will then show the sale value of the annual catch (revenue) at various levels of effort. If each boat-year of effort has the same cost, total costs will be proportional to the amount of effort and may be represented by a linear cost curve starting at the origin.

The net benefit (the excess of revenue over cost) generated in the fishery is in the nature of a resource rent attributable to the quality of the resource as reflected in its attractiveness to consumers and its accessibility to fishermen. The more highly the fish is valued by consumers, the greater the revenue from the catch; the more accessible the stock is to fishermen, the lower the costs of capture. In Figure 2 the maximum net benefit or resource rent that can be generated is *MN*, which occurs at an effort level of *OL*. This position is referred to as maximum economic yield (MEY). Obviously, the resource rent is much

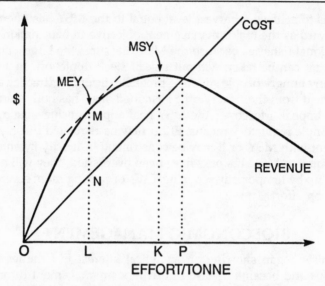

Figure 2. Optimum effort levels for gross value (MSY) and net value (MEY) in a fishery

smaller at MSY. Indeed, with a higher cost curve a loss could be incurred at the MSY level.

It should be noted with respect to the cost curve that economists include in cost a normal return for capital and labor. Thus the resource rent represents a return beyond the profits and wages necessary to keep capital and labor employed at normal levels of remuneration. Therefore the resource rent can be captured as a tax or license fee without endangering the viability of fishing enterprises. If Pacific island countries want to maximize license fees from foreign tuna-fishing operations in their waters, the tuna fisheries must be managed to generate maximum resource rents, which will then be available for capture as license fees.

In a fishery where access is not limited and no significant fees are charged, the equilibrium level of effort will be *OP* boats per year, the amount of effort at which boats will break even on normal returns. If there were more boats, cost would exceed revenue, and boats would start leaving the fishery. If there were fewer boats, revenue would exceed cost, and above-normal profits would be made in the form of resource rent shared by the boats in the fishery. Under open-access conditions, additional boats would be attracted by these profits until effort reached the equilibrium level of *OP* boats.

As the analysis indicates, there are no net benefits in the fishery when *OP* boats participate per year; hence there is no room for governments to generate revenue from the fishery. At the lower effort level of *OL* boats per year, maximum resource rent is generated, which may be left for the owners of fishing boats to enjoy or be captured by government in whole or in part through taxes and license fees. It is possible, of course, to discriminate among different groups of vessels in the fishery. For instance, as a matter of national policy, a country may allow its own vessels to retain some of the rent but impose heavier taxes or fees on foreign vessels so that they retain little or no rent.

If governments impose taxes or license fees equal to *MN* on a fleet of *OP* boats, they will tend to drive out a large part of the fleet until only *OL* boats are left. At this point, fleet revenue (*ML*) will be sufficient to cover *MN* in taxes and fees as well as *NL* in costs, so that equilibrium in the fleet can be restored. Driving away a large part of the fleet would be a painful process, avoidable by limiting access from the start to *OL* vessels only.

One of the simplifications of the foregoing analysis is an assumption that any level of output from the fishery can be sold at the same price. This assumption may be realistic in a small fishery whose catch is disposed of in a large market with the output of other fisheries. In such a case, any variation in the output of the small fishery would be insignificant in relation to the large market and would therefore not influence the market price. But this situation does not prevail in the tuna fishery of this region. Given the importance of the tuna catch of the Pacific islands region to the world market, substantial variations in that catch can have an observable impact on world tuna prices. Any realistic economic model of the Pacific islands region tuna fishery must therefore include a sloping demand curve signifying the market prices that will prevail at different output levels, other factors being equal.

An appropriate model provided by Copes (1970) is shown in Figure 3. It is a supply-and-demand model, using a conventional price-output diagram. It features a supply curve for an open-access fishery, estimated from the yield-effort and cost data portrayed in Figure 2. The supply curve is backward-bending because of the natural upper limit to the catch (*OA*) conforming to MSY. Assuming equal costs per unit of catch for all vessels, the supply curve is also an average cost curve (AC). The conventional sloping demand curve represents average revenue (AR). Under free-entry conditions, the fishery will be in equilibrium when supply and demand are balanced at point *B* with an output of *OC*.

Optimization of economic returns in the situation portrayed in Figure 3 is not as straightforward as in the case of Figure 2, where an unambiguous MEY could be identified. In the case of Figure 3, two kinds of net economic benefits can be generated. First there are the resource rents, discussed in the context of Figure 2. In the Figure 3 sit-

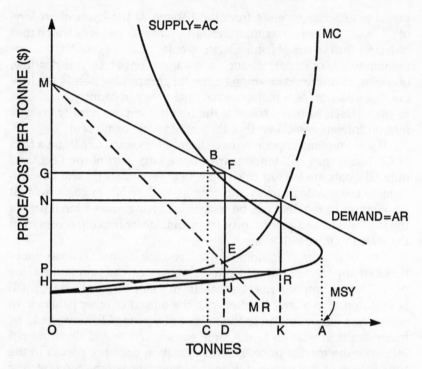

Figure 3. Price and output levels yielding maximum economic benefits
in a fishery

uation, with a sloping demand curve, there is also the possibility of
generating a consumers' surplus consisting of the difference between
what a product is worth to consumers (the highest price they would
pay if the product were available only at such a price) and what they
actually have to pay.[6] The larger the supply and the lower the (equilibrium) market price, the larger the consumers' surplus. However, to obtain the largest resource rent, effort and output must be restricted to
a decidedly lower level. Maximization of consumers' surplus thus requires a distinctly larger output than maximization of resource rent does.
If both forms of net benefit are considered equally valuable, the combination of consumers' surplus and resource rent could be maximized
at an optimum intermediate level of output.

Given this model, what would be the logical position for a management authority attempting to maximize net economic benefits for the
Pacific islands countries by managing effort and output levels in the
tuna fishery?[7] As almost none of the tuna catch is consumed in the
region, the consumers' surplus is of virtually no benefit to the islanders.
Therefore it would appear appropriate for the island countries to seek

maximization of the resource rent only, which they could then capture through license fees and taxes. Technically, the output level (*OD*) at which resource rent is maximized is found by ascertaining the intersection (*E*) of the marginal revenue curve (MR) and the marginal cost curve (MC) (Copes 1972a). The maximum resource rent here is measured by the rectangle *FGHJ*.

If Pacific island countries ever become consumers of a large share of the tuna catch in their waters, a case might be made for a management regime that would maximize the sum of resource rent and consumers' surplus, because both components would then benefit the region's population directly or indirectly. The appropriate output level for joint maximization of resource rent and consumers' surplus is *OK*, which is found by ascertaining the intersection (*L*) of the marginal cost (MC) and average revenue (AR) curves (Copes 1972a). This provides a consumers' surplus measured by the triangle *LMN* and a resource rent measured by the rectangle *LNPR*.

The foregoing discussion illustrates the principles by which optimal management positions are calculated to benefit different interest groups. In reality, of course, more complex calculations may be needed—for instance, when consumers' surpluses or rents are to be split between Pacific island countries and others. Indeed, a number of additional considerations put forward in the economics literature may be taken into account. One of these concerns an additional net benefit, a producers' surplus or intramarginal rent that accrues to fishing enterprises that are more efficient than others (Copes 1972a). Another one concerns the calculation of benefits over time, using appropriate discount rates (Clark and Munro 1975). A detailed discussion of these, however, is beyond the scope of this paper.

INTERNATIONAL TUNA MANAGEMENT

The major tuna stocks in the Pacific islands region are of the skipjack (*Katsuwonus pelamis*) and yellowfin (*Thunnus albacares*) species. The region also has sizable stocks of albacore (*Thunnus alalunga*) and bigeye tuna (*Thunnus obesus*) and smaller stocks of other tuna species. The Pacific Ocean accounts for about two-thirds of the world's tuna catch, and the islands region is the most important production area in the Pacific. It accounts for 25 to 35 percent of the world catch (Doulman 1986a, Clark 1986), including an even greater share of skipjack, the most abundant of the tuna species. The 200–mile exclusive economic zones (EEZs) and declared fishing zones (DFZs) claimed by the island countries and territories between 1977 and 1984 contain probably well over 90 percent of the world's commercially exploited marine fish resources.

Articles of the 1982 United Nations Convention on the Law of the Sea (UNCLOS) charge each nation with responsibility for determining the TAC of the resources in its zone, for assessing its own capacity to harvest the resources, for giving other nations access to the surplus, and for setting the terms and conditions (including fees) of access to this surplus. Given the size of the tuna resource, their own limited capacity for exploiting it, and their need for revenue, many of the island countries have opted to sell access rights to DWFNs.

Other UNCLOS articles spell out a requirement for cooperation among all participating states—RANs and DWFNs—in managing the stocks of highly migratory species (including nine species of tuna) for purposes of conservation and optimum utilization. The necessity for international cooperation in managing the migratory species is obvious: if such a stock is to be conserved so as to yield optimum catches, it must be managed at restricted levels of effort. Hence an effective management regime requires that all RANs harboring the stock during its migration and all DWFNs fishing the stock at any time submit to a commonly agreed limit on total effort compatible with the TAC.

To call for such cooperation is one thing; to achieve it is another. The UNCLOS offers no guidelines for structuring a cooperative management regime and no process for compelling agreement. In the case of the tuna resource in the Pacific islands region, four circumstances make it particularly difficult to structure a management regime and achieve agreement on it: (1) the wide range of stock migration, (2) the complexity of stock composition and migration patterns, (3) the large number of RAN and DWFN participants, and (4) the fact that many stock components spend considerable time in the waters of the high seas.

The analytical models discussed earlier implicitly assumed a fishery on a biologically self-contained stock of a single species. But fisheries management conditions are rarely that simple—certainly not in fisheries that exploit a complex of many overlapping stocks of several species.

The simplest approach is to treat a mixed-stock complex as if it were a single undifferentiated stock—sometimes the only practical way of dealing with the management problem. However, a managed exploitation rate appropriate for a strong stock component may be too heavy for a weak component. Regulating fishing effort to target more heavily on stronger components and less heavily on weaker components by setting gear specifications and imposing area closures can help, but with mixed tuna stocks this course achieves imperfect selectivity at best.

Another difficulty is that fishing units have a monetary incentive to target on the most valuable species and sizes, making them particularly vulnerable to overexploitation. A large fishery on yellowfin and skipjack in the eastern Pacific is an example. Because yellowfin is the more valuable species, it is more heavily targeted.[8] The intensive exploitation of yellowfin in the 1960s led to agreement among the principal participant nations to set up a conservation regime for that species through the IATTC. From 1966 to 1979 an annual overall TAC for yellowfin in a defined conservation area was adhered to by IATTC members, and the fishery was closed each year when estimated catches approximated the TAC level.

Fishing on the abundant skipjack in the area was permitted during the closed season for yellowfin. But the skipjack fishery inevitably took incidental catches of yellowfin. This outcome posed a typical mixed-stock management dilemma: Should fishing vessels be allowed to retain incidental catches of yellowfin? Or should they be required to return them to the sea? In the latter case, much of the incidental yellowfin catch would be wasted, because mortality is high among fish dumped after catching. But allowing vessels to keep the yellowfin would give them an incentive to "accidentally" take large incidental catches or even to target directly on yellowfin. A compromise was accepted, allowing retention of a 15 percent incidental catch of yellowfin. This halfway measure probably reduced the wasteful dumping of excess yellowfin catches without eliminating it.

The IATTC was reasonably successful in maintaining a conservation regime for yellowfin tuna in the eastern Pacific until the management agreement among the major participating nations broke down in 1979. Since then, yellowfin stock components have clearly been overfished, while skipjack components appear still to be capable of sustaining larger catches than are being taken. The breakdown of the IATTC regime illustrates the difficulty of achieving and sustaining a management agreement when many nations are involved.

Obstacles to agreement are particularly severe in the tuna fisheries of the Pacific islands region, where stocks traverse the waters of 22 countries and territories (and the high-seas enclaves between them) and where the vessels of a dozen DWFNs already fish the stocks. Furthermore, there are migratory stock exchanges between the waters of states in the Pacific islands region and the waters of other Pacific states. The complexity of these circumstances has prompted suggestions that tuna management be coordinated across the whole Pacific, involving even more countries in a management agreement (Castilla and Orrego 1984).

When just two countries fish a transboundary stock, it may not be difficult to design a joint management regime and agree on a straightforward division of benefits that leaves both parties better off than they would be in a destructive attempt to outfish one another (Munro 1979). But achieving a management agreement when many parties are involved is much more difficult. The process of negotiation itself can be formidable, quite apart from the difficulty of achieving consensus when so many competing interests are at stake and when benefits can be divided in innumerable ways. A "fair" compromise is hard to define—and even harder to negotiate—when the contending national representatives hold self-serving notions of what is fair and are driven to demonstrate to their governments and their compatriots that they have bargained effectively. Negotiations can end in stalemate or in a flawed compromise that fails to optimize collective benefits from the fishery, that is difficult to implement, and that is open to disruptive disputes over interpretation and enforcement.

When many participants have independent access to the fishery, the "free rider" problem may become severe. Even if the participants agree to constrain their fishing effort to conform to an overall management plan, a renegade country may elect to stay out or pull out of the agreement and reap the benefits of unrestrained fishing while the insider countries attempt to conserve the stock. If the renegade's catch is a small part of the total catch, insider countries may choose to ignore the transgression. Or they may induce the transgressor to join the agreement by offering favorable conditions of participation. But the larger the number of countries involved in the fishery, the larger will be the number of potential trangressors seeking to gain an advantage by stepping outside the agreement. Then the cost of buying the compliance of outsiders while restraining their own fishing effort can become too large for the remaining insiders, and the management agreement can collapse.

Counterbalancing such obstacles are some circumstances that improve the chances of establishing an effective regime. First there is a community of interest among nations—both DWFNs and RANs—to preserve a healthy and economically productive fishery. The DWFNs need fish at a commercially viable cost; the RANs need public revenues from their prime resource. A cooperative agreement for resource management can give both groups what they need—after the inevitable hard bargaining by each party for the largest possible share of the net benefits yielded by the managed resource.

One crucial question is how large an area must be legally controlled to make a regionally based tuna management system workable. International law already allows Pacific island countries to control access to their EEZs or DFZs. But what of the countries outside the region into whose zones the stocks migrate? Must they be drawn into a management agreement? And what of the DWFNs who fish migrating tuna stocks on the high seas? Must they be induced to join the agreement to make it effective?

Ideally, a management regime would draw in both the countries harboring the stocks in their zones at some time and the DWFNs fishing the stocks at any time. Although the obstacles make it almost impossible to establish an ideal regime, there are good reasons to believe that a workable regime can be established on a much narrower basis. What needs to be recognized is that the region's tuna stocks can be fished far more effectively in some places than in others. Waters where stocks tend to concentrate on their migration are much better interception grounds than others; that is, they yield a much higher catch per unit of effort (CPUE), which translates into a much lower cost per unit of catch. This fact is crucially important to the economic viability of fishing enterprises. Vessels are bound to be attracted to the grounds that yield the highest CPUE and to be prepared to pay a price for the privilege of fishing there.

All the signs indicate that the best interception grounds for the bulk of the stocks in the south-central Pacific lie in the zones of some of the island countries. The strongest evidence is the fact that DWFNs are already paying substantial fees to fish in the islands region when they could fish free in the adjacent waters of the high seas. The best interception grounds of all appear to lie in the northwestern section of the region. The seven states in this section—Federated States of Micronesia, Kiribati, the Marshall Islands, Nauru, Palau, Papua New Guinea, and Solomon Islands—have in fact been the most active in concluding access agreements with DWFNs in the effort to generate revenues from their resources. In a move to avoid being played off one against another, these seven states, all members of the FFA, have banded together as the Nauru Group in order to coordinate and harmonize their terms of access (Doulman 1986a, 1986b).

A MANAGEMENT REGIME
FOR THE PACIFIC ISLANDS REGION

Some elements of a tuna management regime in the Pacific islands region are already in place. The FFA is an obvious vehicle for the Nauru Group and other self-governing countries to use for setting up a management regime. (The FFA membership does not include the dependencies of metropolitan powers in the region, but their waters appear to be less productive for the tuna fishery in any case.) The Nauru Group by itself probably offers an adequate geographical basis for a workable tuna management regime because of the high concentration of stocks in its members' waters, but an area that includes the waters of other FFA members would be even better.[9]

Neither the Nauru Group nor the FFA has the data base and the administrative apparatus needed to bring in a full-scale resource management regime. But tuna stocks in the region are not yet under severe pressure, so that there is time to develop a management capability in anticipation of the need for it and at a pace that financial resources will allow.

Once the tuna stocks become more heavily exploited, the most important function of the management regime will be to restrain fishing effort to the optimum level for generating benefits to the Pacific islands region countries—benefits that will be measured by the resource rents generated, at least as long as the fishery is conducted primarily by DWFNs willing and able to pay access fees in relation to the rents their operations generate.

Pacific island countries over time will probably want to increase the share of the tuna catch taken by locally based fleets to generate additional employment and other benefits in the harvesting, handling, and processing sectors, as well as multiplier benefits in linked industries. Doing this may require recalculating the level of exploitation at which their benefits will be maximized when benefits are no longer measured in resource rents. Many of the expanded domestic industrial activities in the region will correlate with the scale of these activities, so that the optimum level of tuna exploitation is likely to be higher than when resource rents alone count as benefits.

In the development of an industrial sector there is a natural process of succession among nations. When a sector depends on advanced technology, the industry is dominated at first by countries able to develop and use the technology. The introduction of sophisticated purse seining technology and techniques helped the United States in the 1960s to become the overwhelmingly dominant tuna-fishing power in the eastern Pacific. Subsequently, Japan reinforced the dominance of its tuna longline fleet in the western Pacific by adding a powerful tuna seine fleet. However, when countries with lower labor costs acquire the same technology and learn skill in its use, they are likely to gain a competitive economic advantage over the original leading nations with their higher labor costs.

Many low-cost countries have acquired skill during the past decade in fishing for tuna with sophisticated purse seiners. In the eastern Pacific they have been displacing the U.S. fleet. IATTC reports show that in its regulatory area the proportion of the tuna catch taken by U.S. vessels dropped from 79.3 percent in 1970 to 57.8 percent in 1978. The subsequent imposition of access fees in some of the waters used by the fleet helped to push down the U.S. share to 34.1 percent in 1985. The purse seine fleets of Latin American states are now taking the larger share of the catch, Mexico in particular having become a major producer of tuna. In the western Pacific such low-cost countries as South Korea,

Taiwan, and the Philippines are beginning to cut into Japanese and U.S. dominance in the tuna fishery, even though longlining for larger tuna remains as important as seining (Doulman 1986a).

A start has been made with the development of a domestic tuna fleet in the islands region, which now accounts for 10 to 15 percent of the region's catch (Doulman 1985, 1986a). Much of this fleet is owned and partially crewed by nationals from outside the region, but there is little reason to doubt that the island countries will in time develop the capacity to dominate tuna harvesting in their own waters. They may be able to use the more labor-intensive technology of pole-and-line vessels to good advantage, at least in localities where surface-dwelling tuna stocks can be matched with local supplies of baitfish and low-cost labor. Moreover, the island states may use their control over a large part of the tuna stocks to give their domestic fleets an advantage in access to the resource. However, U.S. and Japanese control of the major tuna markets may serve as a countervailing power that will induce compromises and retard the islanders' takeover of harvesting operations in their waters.

Developing a cohesive and effective tuna management system in the Pacific islands region or a major part of it will require that a firm collective agreement be reached in four key areas: (1) philosophy of management, (2) fee structure and distribution, (3) quota allocation, and (4) surveillance and enforcement.

The management philosophy must define for whose benefit the management regime should be structured. This, of course, will be linked to the identity of the countries underwriting what might be called the Pacific Islands Tuna Authority (PITA). It is assumed that PITA affiliates would consist of the FFA membership; indeed, the PITA might be organized as a major part of the agency's work. Or possibly it could be organized more effectively as a body serving the needs of a sub-component such as the Nauru Group. Either the larger or the smaller grouping would seem to meet three practical requirements: First, either one is sufficiently sizable and contiguous in marine territory for its joint zone to harbor so large a concentration of tuna stocks that, despite their migration, these stocks cannot be substantially exploited except through access to the joint zone. Second, neither grouping is so large and dispersed that a common management regime would become too complex, cumbersome, and disjointed. Finally, the members of either grouping appear to have sufficient commonality of needs and interests to maintain solidarity and discipline in structuring and enforcing a joint management regime. They are all developing countries that need to extract as much benefit as they can from one of their few abundant resources.

The analysis earlier in this paper suggests that where DWFN fishing is concerned, the PITA will find it advantageous to charge fees that will come close to absorbing the maximum resource rent that the fishery will yield while leaving the operations of DWFN vessels profitable in PITA waters. Finding the ideal fee structure for this purpose will not be easy. Obviously, the fee for each vessel or group of vessels given access should bear a relation to the size of the catch they will take and thus to the amount of resource rent they will absorb. An equally important consideration—one that is often overlooked—is to graduate fees in relation to the productivity of the grounds or substocks that are exploited (Copes 1977). The richer concentrations of fish will yield higher net returns (rents) and be overexploited unless higher fees are charged to fish them; the less rich concentrations will be underexploited unless access to them costs less. Because the richness of fishing grounds within the PITA area varies considerably, a variable fee structure should be considered for practical reasons.

An appropriate way of dividing the fees collected from DWFN vessels is not easily determined. A simple approach would be to let each member country set and collect its own fees—the current practice— even if there has been coordination in setting uniform access conditions. But such a system is almost bound to deteriorate eventually, leading to competition among members to attract larger numbers of DWFN vessels by offering cut-rate fees or other favorable conditions of access— for example, fleet servicing and base facilities.

An agreement among members to charge standard fees and offer standard access conditions while allowing individual members to keep the fees they collect could also break down. DWFN vessels would concentrate their efforts in the most productive waters, and member countries with less productive grounds would collect little or nothing in fees. They might then be induced to break ranks and attempt to draw DWFN vessels with cut-rate fees. Even a fee structure graduated according to the estimated productivity of grounds might yield an unsatisfactory distribution for some less-fortunate members and provoke them to pull out. To maintain solidarity and a strong bargaining position vis-à-vis DWFNs, members might have to reach a political agreement regarding a "fair" distribution of fees. Such an agreement would have to allocate larger benefits to the members with the best fishing grounds in recognition of their right to profit from their superior stocks and to remove the temptation to go it alone. But it would also have to offer enough to the less-favored members to dissuade them from breaking ranks.

Restraints on effort and total catches of tuna in the area generally do not appear to be necessary yet. But the high value of tuna and the trend to improving fishing efficiency suggest that the time may be coming when effort and catch management will be desirable. As was shown above, the highest resource rent—and therefore the highest access fee total—may be obtained when catch and effort are held at an optimal level. Obviously, the research capability and data base of the FFA or an eventual PITA will have to be much improved before optimum levels and patterns of catch and effort can be determined.

When restraints on catch and effort are introduced, the question of catch allocations through quotas becomes important. As PITA countries expand their own fishing capacity, they will want to reserve for themselves a large enough catch quota to meet the needs of their own fleets. To maintain a fair distribution of benefits among them, those that take up quotas should have their share of DWFN access fees reduced accordingly. An appropriate reduction might be an amount equal to what a member's quota allocation would have earned in fees had it been given to a DWFN. After members had taken up all the quotas they wanted, the remainder of the TAC would be allocated to DWFNs. One way of allocating the available quotas among DWFNs would be to auction them off, thereby securing the highest possible revenue from their sale.

When stocks are so highly fished that TAC setting and quota allocation become necessary, it may become desirable to try to harmonize tuna management in the PITA area and in adjacent areas with which there are migration exchanges. These adjacent areas include both the waters of neighboring states and stretches of the high seas. The high-seas interests presumably should be represented primarily by the DWFNs fishing there, though unlimited legal access and absence of police power on the high seas may make the free-rider problem difficult. What is needed is an agreement on harmonizing the tuna TACs of all areas so that the joint TAC will be optimal. Without agreement, parties in each area will be prone to overfish in order to minimize the loss of migrating fish to other areas.

One troublesome problem remains for discussion: surveillance and enforcement (Sutinen and Andersen 1985). The Pacific islands region is an enormous area of over 30 million square kilometers. The zones of FFA-affiliated countries take up two-thirds of this area, including most of the best tuna fishing grounds (Doulman 1986a). Surveillance and enforcement of fishing regulations can be very costly (Lepiz and Sutinen 1985), particularly in such a large area. PITA countries have very limited physical and financial resources for assuming these activities, which must be conducted as economically as possible so as not to absorb too much of the access fees collected. For cost-effectiveness, they will obviously need to vest their surveillance and enforcement authority in a joint service, possibly drawing on the assistance of sympathetic neighboring countries with stronger capacities.

The advent of a new Law of the Sea extending the marine jurisdiction of coastal states has been a fortuitous development for the Pacific island nations. At a time when most of them have gained political independence, they have also gained a measure of economic independence. As most of them possess little else in natural resources, it behooves them to husband well their fishery resources, of which the tuna stocks are the most important. Optimal tuna fisheries management, then, has become a most important matter, calling for an immediate investment and rapid deployment of human and financial resources to give the Pacific islands a strong management capacity at the earliest opportunity.

NOTES

1. The research for this paper has benefited from a grant from the Social Sciences and Humanities Research Council of Canada and from research assistance provided by C. S. Wright.

2. There are two more members of the South Pacific Forum, Australia and New Zealand. They are geographically adjacent to the Pacific islands region. Because in some respects they occupy positions as "metropolitan powers" in the Pacific islands, their interests differ from those of the Pacific island countries.

3. For a full textbook treatment of the fisheries problem, see Cunningham, Dunn, and Whitmarsh 1985.

4. Effort may be thought of, for instance, in terms of boat-years, each unit of effort representing the effect of a standard boat fishing for one year.

5. Although it was an economist (Gordon 1954) who provided the initial analysis of the revenue-cost relationship, it was a biologist (Schaefer 1957) who formulated the basic model represented in Figure 2. It is often referred to as the Gordon-Schaefer model.

6. In reality, of course, most of the first-hand buyers of fish are processors or dealers, not "consumers" in the ordinary sense of the term. However, they represent consumers indirectly. What they are prepared to pay for fish at various levels of output depends on the price they can sell it for, which reflects its worth to the ultimate consumers. Use of the consumers' surplus concept is therefore valid here.

7. A useful and comprehensive review of international tuna management problems, with particular mention of IATTC management efforts in the eastern Pacific, may be found in Joseph and Greenough 1979. For a valuable update, see Joseph 1983.

8. A regulation system allowing full retention of incidental yellowfin catches but imposing penalty clauses to discourage targeting on yellowfin has been proposed by Copes (1972b). Its potential effectiveness is untested.

9. For additional discussion of tuna management options in the Pacific, see Kearney 1977, Kent 1980, and Van Dyke and Heftel 1981.

REFERENCES

Beverton, R. J. H., and S. J. Holt

1957   On the dynamics of exploited fish populations. Ministry of Agriculture, Fisheries, and Food. London.

Castilla, Juan Carlos, and Francisco Orrego Vicuña

1984   Highly migratory species and the coordination of fishery policies within certain exclusive economic zones: The South Pacific. Ocean Management 9:21–33.

Clark, Colin W.

1985   Bioeconomic modeling and fisheries management. John Wiley and Sons. New York.

Clark, Colin W., and Gordon R. Munro

1975   The economics of fishing and modern capital theory: A simplified approach. Journal of Environmental Economics and Management 2(3):92–106.

Clark, Les

1986   Tuna industry developments in the southwest Pacific. In Proceedings of the INFOFISH Tuna Trade Conference. Bangkok. pp. 151–159.

Copes, Parzival

1970   The backward-bending supply curve of the fishing industry. Scottish Journal of Political Economy 17(1):69–77.

1972a   Factor rents, sole ownership, and the optimum level of fisheries exploitation. The Manchester School 40(2): 145–163.

1972b   Resource management in the yellowfin tuna fishery of the eastern Pacific Ocean: Some proposals for the improvement of economic returns through the limitation of fishing effort. Canadian Fisheries Service. Ottawa.

1977   Instituting a management regime for the prawn fishery
              of the Northern Territory of Australia. In Economic im-
              pacts of extended fisheries jurisdiction. Edited by Lee G.
              Anderson. Ann Arbor Science Publishers. Ann Arbor.
              pp. 267–280.

Cunningham, Stephen, Michael R. Dunn, and David Whitmarsh

1985   Fisheries economics: An introduction. St. Martin's Press.
              New York.

Doulman, David J.

1985   Recent developments in the tuna industry in the Pacific
              Islands Region. Pacific Islands Development Program.
              East-West Center. Honolulu. 9 pp.

1986a   Fishing for tuna: The operation of distant-water fleets in
              the Pacific islands region. Research report series no. 3.
              Pacific Islands Development Program. East-West Center.
              Honolulu.

1986b   Licensing distant-water tuna fleets: The experience of
              Papua New Guinea. Marine Policy 11(1):16–28.

Fox, William W., Jr.

1970   An exponential surplus-yield model for optimizing ex-
              ploited fish populations. Transactions of the American
              Fisheries Society 90(1):80–88.

Gordon, H. Scott

1954   The economic theory of a common property resource: The
              fishery. Journal of Political Economy 62(2):124–142.

IATTC

1984   Annual report of the Inter-American Tropical Tuna Com-
              mission 1983. La Jolla, California.

Joseph, James

1983   International tuna management revisited. In Global fish-
              eries perspectives for the 1980s. Edited by Brian J. Roth-
              schild. Springer-Verlag. New York. pp. 123–150.

Joseph, James, and Joseph W. Greenough
1979   International management of tuna, porpoise, and billfish. University of Washington Press. Seattle.

Lepiz, Luis Guillermo, and Jon G. Sutinen
1985   Surveillance and enforcement operations in the Costa Rican tuna fishery. Marine Policy 9(4):310–321.

Kearney, R. E.
1977   The Law of the Sea and regional fisheries policy. Occasional paper no. 2. South Pacific Commission. Noumea, New Caledonia.

Kent, George
1980   The politics of Pacific Island fisheries. Westview Press. Boulder, Colorado.

Munro, Gordon R.
1979   The optimal management of transboundary renewable resources. Canadian Journal of Economics 12(3):355–376.

Schaefer, Milner B.
1954   Some aspects of the dynamics of populations important to the management of commercial marine fisheries. Bulletin of the Inter-American Tropical Tuna Commission 1(2):27–56.
1957   Some considerations of population dynamics and economics in relation to the management of the commercial marine fisheries. Journal of the Fisheries Research Board of Canada 14(5):669–681.

Sutinen, Jon G., and Peder Andersen
1986   The economics of fisheries law enforcement. Land Economics 61(4):387–397.

Van Dyke, Jon, and Susan Heftel
1981   Tuna management in the Pacific: An analysis of the South Pacific Forum Fisheries Agency. University of Hawaii Law Review 3:1–65.

2.

Tuna and the Impact of the Law of the Sea

Anthony J. Slatyer

INTRODUCTION

The Law of the Sea is the "international law" that governs the behavior of nations in the oceans.[1] The Law of the Sea, so far as it relates to tuna in the Pacific islands region, comprises international law in customary and treaty form.

The most recognizable customary international Law of the Sea is that described in the United Nations Convention on the Law of the Sea (UNCLOS), which, while not yet in force and not widely ratified, or even signed, by Pacific island nations, has been endorsed for signature by the South Pacific Forum. This endorsement means that the fisheries management policies of any country, if implemented in accordance with the provisions of the UNCLOS and incorporated in national laws, cannot be lawfully undermined by the fishing activities of foreign fleets. Without these threshold Law of the Sea principles, the Pacific island nations would have no recognized authority to control the activities of foreign vessels fishing beyond their own territorial seas.

Two treaties now in force take full advantage of these principles: the South Pacific Forum Fisheries Agency Convention (FFA Convention) and the Nauru Agreement Concerning Cooperation in the Management of Fisheries of Common Interest (Nauru Agreement). Pacific island nations have concluded many other agreements with nations whose people fish for tunas in the region.

Tunas of all kinds are highly dispersed; in other words, the same species may be found across a wide area of the ocean. Some tunas are also highly migratory, spending their lives crossing vast expanses of the ocean. These characteristics of tunas cause the fishery to come within the scope of many rules of the Law of the Sea.

Because tunas live within the territorial seas, internal waters, and archipelagic waters of countries, they come within the scope of the rules of customary international law described in Parts 2 and 4 of the UNCLOS. Because tunas live within the 200-mile fishing zones of countries, they come within the scope of the FFA Convention, the Nauru

Agreement, and the rules described in Part 5 of the UNCLOS as to living natural resources within the exclusive economic zone (EEZ) of any nation. Because tunas live within high-seas areas, duties are imposed on fishing nations by the principles described in Part 8 of the UNCLOS. Because tunas are a shared stock (living within the fishing zones of neighboring countries), a straddling stock (living in the fishing zone of one country and in adjacent areas of the high seas), and a highly migratory species, the various duties of international cooperation described in Part 5 of the UNCLOS apply.

TERRITORIAL SEAS AND INTERNAL AND ARCHIPELAGIC WATERS

The UNCLOS has not altered Pacific island nations' rights to manage tuna fisheries within their territorial seas and internal waters. As a result of the development of the Convention, it is now accepted that the territorial sea should not exceed 12 nautical miles in breadth (UNCLOS Article 3).

The existence of "archipelagic waters" jurisdiction (for countries comprising closely grouped islands) is now clearly recognized by customary international law (UNCLOS Article 49.1). Several Pacific island nations have archipelagic status.[2] A significant aspect of the confirmation of the archipelagic waters regime has been the recognition that fisheries within archipelagic waters, previously regarded by fishing nations as high-seas fisheries, are within the full control of the archipelagic state. As a result, Pacific island nations now have the unqualified right to regulate the harvesting of the prolific tuna stocks commonly found within their archipelagic waters (UNCLOS Article 49.2).

EXCLUSIVE ECONOMIC ZONES

In the Pacific islands region, the 200-mile fishing zone is the main element of the EEZ regime created by the UNCLOS. Without this regime, each Pacific island nation would have to rely on a unilateral assertion of jurisdiction in any attempt to regulate foreign fishing beyond its territorial sea; in doing so it would directly confront the strength of the fishing nations. (A historic confrontation of this nature was the "Cod Wars" between Iceland and the United Kingdom.) The basic rule is now well established that for all maritime resources within their 200-mile zones, coastal states have "sovereign rights" for a wide range of purposes (UNCLOS Articles 56 and 57). Applied to fisheries, the rights extend to "exploring and exploiting" and "conserving and managing" stocks (UNCLOS Article 56.1(a)). There is no tuna fisheries management activity that does not come within the scope of these rights. The rights to manage fisheries are just as wide-ranging in the 200-mile zone as they are in the territorial sea and archipelagic waters.

The right to manage tuna fisheries is the right to make and to enforce laws prohibiting or regulating tuna activities by the fishermen of any country. Provisions in the UNCLOS describe what may be done in the exercise of these "sovereign rights." Foreign fishermen are required to comply with the conservation measures and with the other terms and conditions established in the laws and regulations of the coastal state, which may take "such measures as may be necessary" to ensure compliance with those laws and regulations (UNCLOS Articles 62.4 and 73.1). This Law of the Sea principle underpins rigorous national surveillance and enforcement programs as well as heavy penalties included in national laws to deter illegal fishing.

A practical problem in making good the right to control tuna fishing in the 200-mile fishing zone and in the territorial sea and archipelagic waters is the accepted principle of the Law of the Sea that fishing vessels have a right of free navigation through these waters.[3] In the Pacific islands region, where fishing vessels travel great distances through many fishing zones in search of tuna and where surveillance capability is limited, this right risks undermining the island nations' rights to control fishing. Fishing nations active in the region have accepted that island nations are entitled to require transiting fishing vessels to comply with requirements for reporting and gear stowage in their fishing zones to minimize the risk of their engaging in fishing during transit through those waters. This limitation on the right of transit passage is a necessary corollary to the sovereign right to manage the fisheries resources. The fishing vessel, because it exists only to undertake resource-related activity, must meet obligations that normal vessels need not meet.

Even though the Pacific island nations' rights to manage the tuna resources to their own advantage are absolute, other principles of the Law of the Sea establish a range of duties that must be observed by all coastal states, including the Pacific island nations. In gaining sovereignty over tuna fisheries, the island nations have gained a form of custodianship, with duties to conserve and to use tuna stocks. Principal among these duties is to ensure that fisheries in the 200-mile zone are not "endangered by overexploitation" (UNCLOS Article 62.2). Stocks are to be kept at levels "which can produce the maximum sustainable yield, as qualified by relevant environmental and economic factors, including . . . the economic needs of coastal fishing communities and the special requirements of development States" (UNCLOS Article 62.2). An "allowable catch" must be determined for each fishery, requiring in turn fairly comprehensive data about the size, distribution, and migration habits of the fish (UNCLOS Article 61.1). In the tuna fishery, these obligations are effected by limitations on fishing effort written into access agreements and by detailed reporting requirements generally imposed by national laws. Fishing nations have generally been reluctant to cooperate in supplying data without having a concomitant role in developing fisheries management policies. Such a trade-off is not contemplated by the Law of the Sea.

Beyond the duty to protect tuna fisheries from overexploitation, each Pacific island nation is required to "promote the objective of optimum utilization" of the fisheries in the zone by making available for harvesting by other countries any stocks within the "allowable catch" that it cannot harvest itself (UNCLOS Article 62.2). It has been argued by some fishing nations that this provision in effect guarantees them a share in any underexploited fishery. This approach, which mistakenly translates coastal state "duties" into fishing nation "rights," is not consistent with the Law of the Sea principle that the power to make management decisions about the harvesting of fish in the fishing zone of a country is vested exclusively in that country (UNCLOS Article 56.1(a)). In the Pacific islands region, the Law of the Sea allows each island nation to determine, without influence from any fishing nation, whether there are in fact excess tuna stocks in its fishing zone and when, to which country, and in what manner such stocks should be allocated. That the Pacific island nation has the duty to satisfy itself that these decisions will not lead to the overexploitation or the underutilization of the fishery does not diminish its sovereign right to make the decision, nor does it invest any fishing nation with some "right" of access to excess stocks.

Foreign fishermen are required to comply with the terms and conditions of access established by Pacific island nations (UNCLOS Article 62.4(a)). Generally in the region, significant terms and conditions, such as fees and reporting requirements, are negotiated with fishing nations, not established unilaterally. There is no basis in international law for access conditions to be negotiated. Negotiations can, however, achieve commercial and political objectives that are unattainable unilaterally. For example, in the context of negotiating an access agreement, a fishing nation can be pressed directly to agree to provide other benefits, such as fee supplementation and guarantees of the good behavior of its fishing fleet.

SHARED STOCKS

Because tuna stocks live within the 200-mile zones of neighboring Pacific island nations, the Law of the Sea requires these nations to "seek to agree upon the measures necessary to coordinate and ensure the conservation and development" of the fishery. Such agreement is to be sought either directly or through appropriate subregional or regional organizations (UNCLOS Article 63.1).

Pacific island nations have a long tradition of direct consultation on matters of common interest; shared tuna fisheries are no exception. These consultations are ongoing and, on occasion, formalized. For example, groupings of Pacific island nations have come and gone for the purpose of negotiating access arrangements with U.S. fishing interests.[4] These nations have also reached formal agreement among themselves on how these access arrangements should be administered.

Through organizations at a regional level, all 16 independent and fully self-governing Pacific countries, as parties to the FFA Convention, are members of the South Pacific Forum Fisheries Agency (FFA).[5] They meet regularly through the Forum Fisheries Committee to resolve fisheries issues of common concern. The committee is required under the FFA Convention to promote intraregional coordination and cooperation on a number of matters, including harmonization of policies on fisheries management, collaboration in surveillance and enforcement, and cooperation in respect of access to each other's 200-mile zones (FFA Article 5.2) The members of the agency also have a direct obligation under the FFA Convention to provide the FFA (and therefore each other) with any available and appropriate information, including catch-and-effort statistics, laws and agreements, and biological and statistical data (FFA Article 9).

Through organizations at a subregional level, the seven neighboring island nations in the western Pacific that are parties to the Nauru Agreement have agreed to "coordinate and harmonize the management of fisheries with regard to common stocks within the fisheries zones, for the benefit of their peoples" and to adopt a uniform approach to the activities of foreign fishing vessels.[6] The Nauru Agreement is to be fully implemented through subsidiary arrangements. It has not been necessary so far to introduce any such arrangements for tuna management.

STRADDLING STOCKS

Each Pacific island nation in whose 200-mile zone tuna stocks are found and the fishing nations whose nationals harvest the stocks in adjacent high-seas areas are required to "seek to agree upon the measures necessary for the conservation of these stocks in the adjacent [that is, high-seas] area" (UNCLOS Article 63.2). Because this provision affects only the fishing activities of the fishermen of the fishing nation, the obligation is primarily on the fishing nation to seek the adjacent island nation's views on whether conservation measures are necessary and, if so, what they should be. There has been no evidence yet that any fishing nation is willing to give effect to this requirement.

HIGHLY MIGRATORY STOCKS

Tuna is "highly migratory" for the purposes of the Convention (UN-CLOS Annex I). For this reason the Pacific island nations in whose 200-mile zones the tunas live and the fishing nations whose nationals harvest the stocks are required by Article 64 of the Convention to cooperate for two purposes: "conservation" and "optimum utilization." Securing these outcomes is the responsibility of each Pacific island nation for the stocks within its zone, of the fishing nations for the stocks within the high seas.

Controversy surrounds the meaning of Article 64. Fishing nations desiring to participate in tuna fisheries management in the region have asserted that the article requires cooperation to be effected by an international organization in which the fishing nations are directly involved. Prior to the era of 200-mile fisheries zones, when all commercially important tuna fisheries were high-seas fisheries, organizations representing the interests of each nation whose fishermen were active in the fishery had been formed to establish international tuna management regimes. In the eastern Pacific there was the Inter-American Tropical Tuna Commission (IATTC); in the north Atlantic there was the International Commission for the Conservation of Atlantic Tunas (ICCAT). Any suggestion that the Law of the Sea now requires organizations of this kind to be established or used is answered by the plain words of Article 64, which confer a choice whether to cooperate "directly *or* through an international organization." The requirement to cooperate through an organization arises only when this mode is chosen by the countries concerned.

As an alternative to developing new broad-based tuna management organizations, it has been asserted that fishing nations have a right to participate in the decision-making processes of the island nations. Such a right given to the fishing nations is plainly at odds with the sovereignty given exclusively to the island nations by Article 56. This idea also arose before there was any significant national jurisdiction over tuna stocks. It has no relevance today.

There is nothing to prevent fishing nations from offering to island nations their opinions and concerns about the conservation or utilization of stocks in the fishing zones. An island nation may be disposed to consider those opinions when making management decisions about fishing if the fishing nation agrees to take the island nation's views into account when making decisions about the activities of its fishermen in adjacent high-seas areas. Such a reciprocal undertaking would be consistent with the "direct cooperation" objectives of Article 64.

Another controversial interpretation of Article 64 is the U.S. claim that the sovereignty given by Article 56 does not extend to highly migratory species. This claim is based on (1) the perceived strength of the cooperation requirement described in Article 64 and (2) a view that customary international law, while recognizing extended coastal state jurisdiction over fisheries generally, does not afford the same recognition to some kinds of highly migratory species—namely, tunas. The former basis denies the plain words of Article 64, which is expressly stated to be read in conjunction with Article 56 establishing coastal state sovereignty. The latter basis defies explanation, since the United States is the only country holding that opinion, and one country's view has never sustained customary international law. It is widely accepted that the rationale behind the U.S. position is to strengthen the position of its distant-water tuna industry in negotiating a cost advantage over its Asian competitors.

Article 64 has been effected in the Pacific islands region through direct cooperation via access agreements and joint-venture agreements with many fishing nations and fishing companies. Collectively, these agreements grant foreign fishermen access in varying degrees to the fishing zones of all FFA member states. These agreements reflect a high degree of direct cooperation with all interested fishing nations for the purpose of "optimum utilization" within the meaning of Article 64. If and when conservation measures become necessary or desirable, there is no reason that effort limitations cannot be introduced into those agreements.

HIGH-SEAS STOCKS

Pockets of high seas are enclosed by the 200-mile zones of Pacific island nations; large areas of high seas are adjacent to the area covered by the combined zones of the Pacific island parties. The general principle applicable here is that high-seas fisheries are open to the fishermen of all countries to fish wherever and however they wish, without interference from any other country (UNCLOS Article 116). Two significant exceptions to this principle are relevant to the management of tuna fisheries.

First, freedom of fishing on the high seas does not allow a fishing nation to disregard the interests of any island nation whose domestic fishery could be affected by the amount or type of fishing carried out in adjacent high-seas areas (UNCLOS Article 116(b)).

Second, all nations whose fishermen are active in high-seas fisheries are required to cooperate in applying any necessary conservation and management measures (UNCLOS Article 117). In particular, when fishermen of different countries are fishing the same stocks in the same area, their governments are required to agree upon any measures necessary for the conservation of the fisheries concerned (UNCLOS Article 117). This is a much stronger requirement than the one in Article 64, reflecting the fact that in the high seas no coastal state sovereignty overshadows the international management requirement. Management measures must be nondiscriminatory and based on the best available scientific evidence of what the maximum sustainable yield is, as qualified by relevant environmental and economic factors (UNCLOS Article 118). Populations of the stocks are to be maintained above the levels at which their reproduction may become "seriously threatened" (UNCLOS Article 118).

IMPACTS ON THE RESOURCE

It is axiomatic that tuna, being a highly dispersed fishery fished by many nations, cannot be effectively conserved by the action of any single island nation nor by the restraint of any single fishing nation. When conservation becomes necessary, international effort of some kind will be required.

In the South Pacific, the Law of the Sea has invested the island nations collectively with custodianship over a vast area of tropical tuna grounds. This makes it possible for island nations to manage the resource effectively by developing cohesive management practices and enforcing them against the fishing nations over which they have leverage, either unilaterally or through fishing access agreements. In other words, effective resource management may be possible without the cooperation of the fishing nations. Pacific island nations should, by implementing coordinated fishing effort limitations, be able to conserve stocks while maximizing their own return from the same fishery. Such cooperative action is entirely consistent with the Law of the Sea as ensuring the conservation and development of a shared fishery; it can be swiftly effected under the FFA Convention or the Nauru Agreement should the need arise.

ECONOMIC AND POLITICAL IMPACTS

Pacific island nations now have recognized rights to make and enforce laws to control fishing within their 200-mile zones. The opportunity is theirs to control it in a way that generates benefits to them.

The most significant constraints in realizing these benefits are (1) the ever-present risk of unauthorized fishing activity and (2) the vulnerability of the island nations to being traded off against one another by large and powerful fishing nations seeking to obtain access to tuna resources on the most favorable terms possible. The risk of unauthorized fishing is principally a function of surveillance and enforcement capability; it is clearly exacerbated when access rights are not available. The "trading off" is by means of threatening to deny access fees, which, in the case of the smaller nations, are critical revenues for budgetary purposes. An added burden in overcoming these constraints is the legislatively entrenched policy of the U.S. government not to recognize that the island nations have jurisdiction over the tuna in their zones.

It is a real challenge for Pacific island nations to make good the powers and rights given them by the Law of the Sea so as to obtain the best possible deal in these difficult circumstances. But they are achieving it in several ways—all possible because of the Law of the Sea.

A system of agreements that regulates the amount of fishing in the whole of the South Pacific tuna fishery within national fishing zones empowers Pacific island nations collectively to control access to the whole fishery. Over time, the full coordination of access controls will maximize the benefits available to the island nations from foreign fishing. Uniformly applied effort limitations in such a large fishery will necessarily lead to significant increases in the value of access rights. Similarly, the continued development of a pattern of selective restrictions on particular kinds of fishing will encourage other fishing methods that may offer the island nations better opportunities for employment or participation in the industry.

Substantial progress has been made in securing compliance with national laws through the development of regional sanctions, innovative legislation, effective systems of surveillance and enforcement, and the inclusion in new fishing agreements of provisions that impose significant responsibilities for surveillance and enforcement on the governments of fishing nations. An example of an effective regional sanction is the Regional Register of Fishing Vessels maintained by the FFA. In addition to providing a data base on all fishing effort in the tuna fishery, the register serves as a blacklist: any vessel refusing to submit to the jurisdiction of any participating nation risks being excluded from the zones of all FFA member nations. The sanction is such an effective deterrent that it has never had to be fully invoked.

More recently the island nations have recognized that in selling access rights, they are also selling diplomatic good will and a strategic maritime presence—collateral benefits potentially of more value to fishing nations than the access rights themselves. The value of these benefits is reflected in the higher fishing fees recently negotiated in the region.

No island nation alone can confer these benefits. In all cases, the island nations stand to gain by cooperating with each other. For example, a coordinated approach by Pacific island nations to the licensing of Japanese tuna vessels led directly to substantially increased fee revenues and the development of fee formulas that avoid the need for burdensome annual renegotiations. Similarly, the value of the collateral benefits attaching to the regional license to be sold to U.S. fishermen will far exceed the value of the benefits that island nations could offer separately.

There are also clearly ascertainable savings in cooperating in the use of common data for targeting scarce surveillance and enforcement resources and in using a central agency representing all the island nations to receive raw data and process it into a form useful as a basis for decision making by island governments. In such ways the Pacific island nations, by cooperating among themselves, are making much of the opportunities presented by the Law of the Sea.

So long as these mechanisms are used, the Law of the Sea will have equipped the island nations with the means to deal more equitably with the outside world and the recognized right to manage the tuna resources of the region for their own benefit.

NOTES

1. The views expressed in this paper are the author's own and do not necessarily reflect the views of the Australian government.

2. For example, Papua New Guinea, Solomon Islands, Vanuatu, and Fiji are archipelagic states.

3. See Part 2, Section 3, of the Law of the Sea Convention for navigation through the territorial sea, Articles 52 and 53 for navigation through archipelagic waters, and Articles 58.1, 87, and 90 for navigation through the exclusive economic zone.

4. Federated States of Micronesia, Marshall Islands, and Palau entered into an arrangement with the American Tunaboat Association (ATA) for purse seine fishing under a single license in 1980. Federated States of Micronesia, Kiribati, and Palau entered into a similar arrangement for 1983 and 1984. Tuvalu, Western Samoa, Cook Islands, Niue, and New Zealand (for the Tokelau Islands) entered into a similar arrangement for 1983 and 1984.

5. Australia, Cook Islands, Federated States of Micronesia, Fiji, Kiribati, Nauru, New Zealand, Niue, Palau, Papua New Guinea, Solomon Islands, Tonga, Tuvalu, Vanuatu, and Western Samoa are FFA members. Marshall Islands participates fully in the work of the agency as an observer.

6. Papua New Guinea, Palau, Federated States of Micronesia, Marshall Islands, Kiribati, Nauru, and Solomon Islands are parties to the Nauru Agreement.

REFERENCES

South Pacific Forum Fisheries Agency (FFA)

1979   South Pacific Forum Fisheries Agency Convention. Honiara, Solomon Islands. 6 pp.

United Nations

1983   The Law of the Sea: United Nations Convention on the Law of the Sea. New York. 224 pp.

3.
Biological Perspectives on Future Development of Industrial Tuna Fishing

John Sibert

INTRODUCTION

The fisheries discussed in this book depend principally on stocks of two species of tuna: skipjack (*Katsuwonus pelamis*) and yellowfin (*Thunnus albacares*). Table 1, a summary of tuna fishing data supplied to the South Pacific Commission (SPC) between 1981 and 1985, makes the importance of these two species obvious: yellowfin comprise over 70 percent of the longline catch; skipjack and yellowfin together comprise over 95 percent of the total surface (purse seine and pole-and-line) catch. These two species are caught by the fleets of several nations using three principal gear types: longline, pole-and-line, and purse seine. Yellowfin are caught by all three gear types—predominantly by longline but, more recently, by purse seine as well. Skipjack are caught by both pole-and-line and purse seine.

The purpose of this paper is to review information available on skipjack and yellowfin tuna stocks and information on the three fisheries that exploit them in the waters of the Pacific island states. An attempt is made to draw conclusions about their future potential. In the process, areas of uncertainty are exposed and the implications of these uncertainties explored.[1]

HISTORY AND CURRENT STATUS

Although tuna fisheries are discussed elsewhere in this volume by fishing method, some points are mentioned here in order to set the historical context for interpreting current conditions.

Fishing for tuna in the tropical Pacific is probably as old as human settlement in the region, and artisanal tuna fishermen are found today in almost every island community (Gillett 1985a, 1985b). In the first decades of the twentieth century Japanese fishermen began to extend

Table 1. Annual summaries of catch by gear and by species 1981–1985

Gear and species	1981	1982	1983	1984	1985
Purse seine			Tonnes		
Skipjack	20,954	53,487	77,117	213,264	158,371
Yellowfin	9,522	21,706	20,386	87,638	53,432
Other	292	867	860	2,482	1,455
Total	30,768	76,060	98,363	303,384	213,258
Pole-and-line			Tonnes		
Skipjack	38,385	21,075	45,775	33,014	20,612
Yellowfin	303	1,160	836	736	981
Other	406	377	311	70	194
Total	39,094	22,612	46,922	33,820	21,787
Longline			Pieces		
Albacore	123,125	178,362	125,223	263,590	196,974
Bigeye	211,250	272,098	188,019	322,368	420,417
Yellowfin	1,097,512	994,871	939,996	842,178	929,241
Billfish	51,600	52,890	40,659	75,753	71,313
Other	50,805	30,055	18,273	31,785	34,500
Total	1,534,292	1,528,276	1,312,170	1,535,674	1,652,445

Source: South Pacific Commission.

their activities southward from their home islands; their early efforts initiated the distant-water fishing era. The most substantial expansion of distant-water fishing began in the late 1940s as Japanese fishing fleets resumed operations after the war. This phase was aided by technological improvements in handling bait and catch that facilitated longer trips and increased carrying capacity (Matsuda and Ouchi 1984).

Expansion of tuna fisheries became explosive in the early 1980s when Japanese fishermen introduced purse seining to the region. Although purse seining is widely used in other areas of the world, its introduction into the western Pacific depended on modifications to nets and net-handling equipment and changes in fishing habits. Poor fishing conditions in the eastern Pacific during the 1982–83 El Niño stimulated further expansion as many U.S. purse seine vessels moved west.

Figure 1 shows trends in skipjack and yellowfin catches in the waters of the Pacific island states since 1980 for the three principal gear types. Purse seine catches, negligible in 1979, grew to surpass catches by the

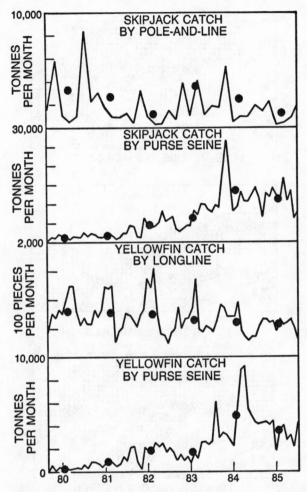

Figure 1. Catch by month for each gear type in the western Pacific since
1980[a]

[a]Circles indicate averages for the year.

other gear types in 1982. In contrast, pole-and-line and longline catches
either declined slightly or remained stable during the same period.

Figure 2 shows the geographic distribution of skipjack and yellowfin catches for each gear in 1983 and 1984, the years for which SPC data
holdings are most complete. Purse seine yields are greatest in the
equatorial region of the western Pacific, that is, in the exclusive economic zones (EEZs) of Papua New Guinea, Palau, and the Federated
States of Micronesia. Pole-and-line and longline fisheries are more
widely dispersed throughout the region. Yields from the longline

42

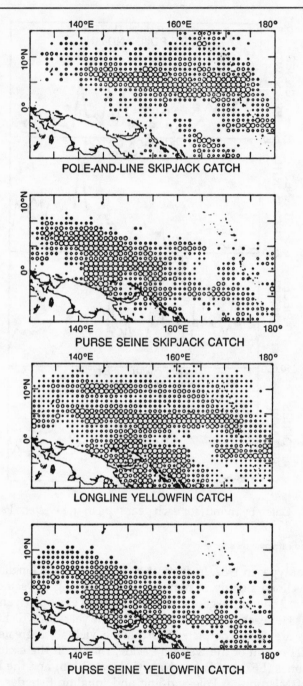

Figure 2. Geographic distribution of catch by gear 1984–1985[a]

[a]The size of the circle is proportional to the total catch for the period.

fishery are greatest in the regions of the north and south equatorial currents and the equatorial countercurrent; they are rather uniformly distributed from west to east. Skipjack yields from the pole-and-line fishery are greatest in the region of the equatorial countercurrent; they are dispersed more toward the east.

Catch per unit of effort (CPUE) is traditionally used as an indicator of stock size by fishery biologists. Figure 3 shows recent trends in both skipjack and yellowfin CPUE for the three fisheries. There are no sharp downward trends that give an unambiguous indication of overfishing. The downward trend in the longline CPUE is a continuation of a long-established decline traceable back to the 1960s.

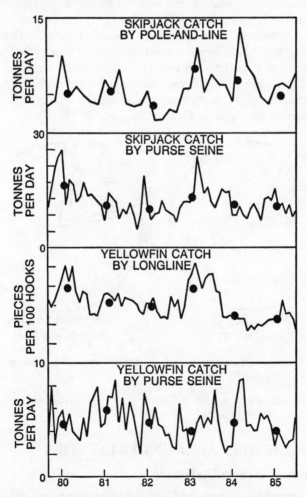

Figure 3. Trends in catch per unit of effort by month for each gear type in the western Pacific since 1980[a]

[a]Circles indicate averages for the year.

A major change accompanying the expansion of purse seine fishery is the dependence on flotsam. Tuna schools are often found in association with something else. In the eastern Pacific there is a strong association between tuna schools and pods of certain species of dolphins—an association exploited by fishermen who hunt for aggregations of dolphins and surround them with their nets in order to catch the tuna school. In the western Pacific an association between dolphins and tuna has not been demonstrated. Instead, tuna schools are found in association with floating logs and other debris. Fishermen hunt for logs rather than for dolphins, often making a single set each day just before dawn.

Dependency on flotsam in the western Pacific is extreme. Few sets are made on freely swimming schools; hence the proportion of the total catch from freely swimming schools is small (Figure 4). The dependency is so great that fishermen mark individual logs with lights and radio beacons so that they can find them and make repeated sets on the same log. Some vessels have taken up the practice of launching artificial flotsam, or *payaos*, to compensate for the lack of logs—a practice common in the Atlantic Ocean (Bard et al. 1985).

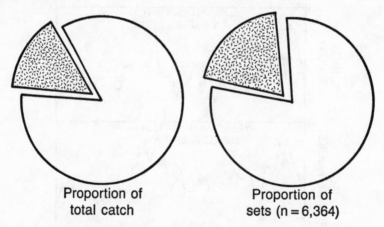

Proportion of
total catch

Proportion of
sets (n = 6,364)

Figure 4. Proportion of catch and set by set type for Japanese purse seine vessels 1985[a]

[a]Shaded areas indicate sets on freely swimming schools. Unshaded areas indicate sets on schools associated with debris.

ISSUES AND UNCERTAINTIES

Interaction between Fisheries

Fisheries interaction is a term used to refer to problems associated with multiple fisheries operating on different scales with different gear types and exploiting widely dispersed stocks and highly mobile fish. Some potential problems are the impacts of purse seining on both longline and pole-and-line fisheries for yellowfin and skipjack, of harvests in one area on harvests in another, and of industrial fishing on artisanal fishing.

Prior to the 1980s the major fleets operated in a way that minimized interactions. Pole-and-line fleets exploited surface-dwelling skipjack schools. Longline fleets exploited deeper-dwelling mature populations of yellowfin. Pole-and-line fleets typically operated in different areas from longline fleets. The introduction of large-scale purse seining eliminated these tidy separations. Catches by purse seiners typically contain a mixture of skipjack and small yellowfin. There are important areas of overlap between the purse seine fishery and both longline and pole-and-line fisheries (Figure 5).

Aspects of interactions between fisheries have been discussed previously by Lenarz and Zweifel (1979), Sibert (1984), and Hilborn (1985b). Because of the large number of vessels, the wide geographic area, and the highly valued product, the longline fishery merits special attention.

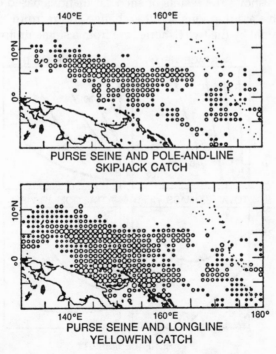

PURSE SEINE AND POLE-AND-LINE
SKIPJACK CATCH

PURSE SEINE AND LONGLINE
YELLOWFIN CATCH

Figure 5. Geographic distribution of overlap in catch by gear 1984–1985[a]

[a]The size of the circle is proportional to the extent of catch by both fleets.

Interaction between purse seine and longline fisheries for yellowfin arguably has the greatest potential for adverse impact.

There are broad geographic areas of overlap between longline and purse seine fisheries (Figure 5). The average size of purse seine-caught yellowfin is much lower (approximately 5 kilograms) than that of longline-caught fish (approximately 25 kilograms). These size differences reflect a difference in age, and the concern over interaction is based on the simple biological assumption that some proportion of young fish vulnerable to the purse seine fleet grow to an age vulnerable to longline fishing.

Knowledge of growth, mortality, intensity of fishing, vulnerability to fishing, recruitment of individuals to vulnerable stocks, and exchange between stocks vulnerable to the different gears would enable biologists to predict the extent of interaction. Unfortunately, none of this information is available for yellowfin stocks in the western tropical Pacific. Nevertheless, calculations can be made by using information from other parts of the world and making some simple assumptions about exchange and recruitment (Hilborn 1985b).

Figure 6 shows the results of such calculations based on the simplest possible assumptions about exchange and recruitment, namely, that fish caught by the two fisheries are from a single uniformly mixed

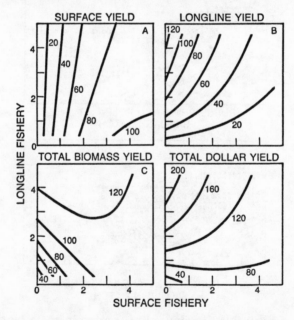

Figure 6. Theoretical yield of combined fisheries at different levels of fishing intensity.

stock and that recruitment to the stock is not affected by fishing. Panel A of Figure 6 gives the calculated yield from the surface fishery in kilograms of fish per individual recruited to the stock in relation to both surface and longline fishing intensities. The surface yield increases in proportion to surface fishing intensity; it is affected by the longline fishery only at very high levels of surface fishing intensity.

Panel B of Figure 6 gives the yield from the longline fishery. Longline yield increases in proportion to longline fishing intensity but is strongly affected by surface fishing at relatively low levels. Panel C of Figure 6 gives the combined yield in kilograms from both fisheries, showing clearly that, in terms of kilograms of fish, a mixture of surface and longline fishing is justified.

Since the price per kilogram of longline-caught fish for the sashimi market is much greater than that of purse seine-caught fish for the cannery market, interaction has important economic consequences. Panel D of Figure 6 illustrates the total yield in dollars from the fishery. At very low levels of longline fishing, economic yield can be increased slightly by introducing surface fishing. However, at higher levels of longline fishing, the surface fishery actually decreases total economic yield.

The obvious question is, Where are current fisheries on these surfaces? The honest answer is that this is not known because there are insufficient data on yellowfin stocks from the western Pacific. The question deserves an answer, however, and on the basis of not being able to detect impacts as severe as predicted at high levels of fishing intensity, it can be concluded tentatively that current fisheries are operating at around two or less (Figure 6). In other words, current fisheries, both surface and longline, are harvesting a small proportion of the stocks. Purse seine activity is having little impact on the longline fishery, but increases in purse seine effort should be monitored carefully.

There are obvious uncertainties in these calculations, and the appearance of the yield curves depends greatly on assumptions about stock mixing and recruitment. Nevertheless, the general conclusions are unchanged: although total biomass yield may be increased by combining the two harvesting methods, economic yield is greatest when surface fishing intensity is low. This conclusion reaches beyond simple biological assessments, but it indicates clearly the importance of the interplay between biological and economic processes. The implications for a manager depend on his point of view, that is, whether he owns purse seiners or longliners or is attempting to maximize benefits from his exclusive economic zone (EEZ).

Effects of Flotsam

The dependency on flotsam is a relatively recent development, one tied conceptually to the use of moored fish aggregation devices (FADs).

There has not been sufficient experience with this fishing practice to build an understanding of its implications for tuna stocks. What little information there is suggests caution. An obvious advantage of flotsam and FADs to fishermen is that they reduce search time—and hence costs—to such an extent as to alter the economic structure of the fishery. Because flotsam and FADs make it possible to harvest fish at a lower margin, it is economically possible for fishermen to continue fishing overexploited stocks and cause further serious depletion (Floyd and Pauly 1984).

Another complication is the possibility of decoupling the fisheries scientist's most useful index of stock size, CPUE, from the stock. In the case of a purse seine fishery operating on free-floating flotsam such as logs, catch per set merely indicates the abundance of logs (assuming a different log each day). Neither of these indicators carries any information on the larger stock of fish from which the flotsam attracts schools.

Some useful calculations are possible, however, if sufficiently detailed data are available. When vessels make repeated sets around the same object, it is possible to calculate the recruitment rate to the object and the size of the larger population from which recruitment occurs (Hallier 1985; Ianelli 1986). Although neither of these calculations truly measures the stock as a whole, both are useful tools for small-scale management. Also, the collection of size information from free-floating flotsam fisheries will serve as an important warning if stocks begin to be overexploited.

High Variability in Yellowfin CPUE

Another way to look at the information in Figure 3 is to examine catch in relation to the amount of fishing effort applied, as shown in Figure 7. In a fully exploited stock these figures show a humped form, with lower yields at higher efforts. The contrast between the plots for skipjack and yellowfin is striking. Month-to-month changes in skipjack catch are generally in proportion to month-to-month changes in effort (that is, along the diagonal). Month-to-month changes in yellowfin catch, on the other hand, are often not in proportion to effort (that is, along the vertical). The dramatic between-month drops in catch at constant effort would lead to the absurd conclusion that catches of a few thousand tonnes result in a halving of the population if one holds to the assumption that CPUE is proportional to stock size. Nevertheless, annual catch is proportional to the average effort, with no indication of low catch at higher efforts—indications that the stock is not seriously depleted.

The causes of this variability are unknown, but there are several related possibilities. The dependence on flotsam may have decoupled

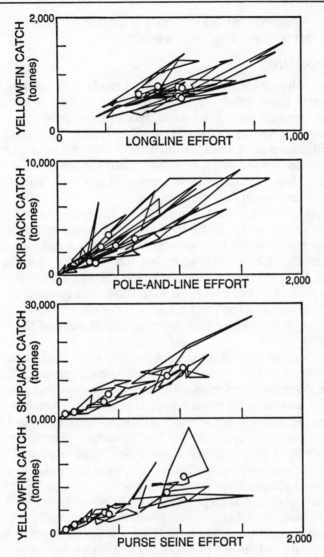

Figure 7. Catch as a function of effort by month for the period since 1980[a]

[a]Circles indicate annual averages.

CPUE from stock size, as suggested above. Under some conditions, purse seine vessels operate in a coordinated fashion, so that small areas may be subject to unusually intense fishing effort. As a consequence, the depletion of the yellowfin stock may be localized and transitory. Either of these possibilities could account for the variations shown in

Figure 7; it is possible that this variability is the first sign that yellowfin stocks are being affected by exploitation.

Data Reporting

Times series data of catch and effort are the fundamental data of fisheries science. These data, often augmented by auxiliary information about size distributions of fish caught, relationships between age and length, tag recapture results, and geographic distribution of catch, are used to calculate potential yields from fisheries. Ultimately, most of the data are obtained from the fishery itself. Finding ways to accomplish the efficient flow of data from fishermen to resource analysts and managers is an eternal fisheries problem.

Data for analyzing western tropical Pacific tuna fisheries come from two essential sources. The first is the historical fisheries data published by various distant-water fishing nations (Fisheries Agency of Japan 1962–1980a, 1962–1980b; Fisheries Research and Development Agency 1980–1985; Tuna Research Center 1973–1983). The practice of publishing fisheries data was discontinued in the late 1970s by some distant-water fishing nations. The second source is the information supplied by fishermen in compliance with licensing requirements of the Pacific island states.

Evaluations of the current state of the fishery depend heavily on the reliability of data from this second source—daily records of fishing activity in the region. Unfortunately, there is no compelling reason to believe that these data are either complete or reliable. Estimates of the percentage of catch reported to regional fisheries agencies ranged from less than 40 percent for pole-and-line fleets to about 90 percent for some purse seine fleets (South Pacific Commission Tuna Program 1986).

Normally a sample size of 40 to 90 percent of a population would be considered adequate; however, there is some reason to believe that the sample is biased. A large proportion of the nonreported catch may be taken in international waters, where reporting is generally not mandatory. Some data (summarized in Table 2) are available, however, which enable comparison between purse seine catches reported without positions (presumbably from international waters) with catches reported with positions (that is, from waters within EEZs). It is clear from Table 2 that fishing conditions were quite different for reported catches with and without positions. The percentage of successful sets was 83 percent in areas where positions were not reported; it was 69 percent in other areas. The result is that CPUE, measured as average catch per day, was 34 percent higher in regions where positions were not reported. Thus there is ample reason to conclude that the sample of the fishery on which management decisions are based is badly biased.

Table 2. Summary of purse seine data reported to the South Pacific
Commission by U.S. flag vessels 1984

	With positions (EEZs)	Without positions (international waters)
Total annual catch	42%	58%
Average catch/day fishing	14.6 tonnes	19.6 tonnes
Set success rate	68.8%	82.9%
Average catch per successful set	29.5 tonnes	30.5 tonnes

CONCLUSIONS

Tuna fishing in the tropical western Pacific has a long history, with several well-established and successful fishing methods. The recent introduction of purse seining to the region has raised uncertainties about how the stocks will be affected and about how to best manage the fishery. The growing aspirations of Pacific island states to obtain a fair share of the revenues from the pelagic fisheries of their EEZs has added to the uncertainties and given a note of urgency to their resolution.

Fundamental facts of tuna biology in this region are simply unknown. The information on which biologists base their analyses is not only incomplete but also biased. Furthermore, economic and political changes have reduced the flow of information from fishermen to biologists.

In spite of these uncertainties, the prognosis for further expansion of tuna fishing is good. There are few obvious causes for alarm with respect to general levels of exploitation. The uncertainties, intensified by the great importance of this resource to the continued economic development of the countries of the region, argue for caution in planning for expansion and provide impetus for more thorough data collection and more extensive research.

NOTE

1. This contribution is a synthesis of work by several members of the South Pacific Commission Tuna and Billfish Assessment Program. In particular, I would like to thank Richard Farman, Ray Hilborn, James Ianelli, and Tom Polacheck for their invaluable input.

REFERENCES

Bard, F. X., J. M. Stretta, and M. Slepouka

1985   Les épaves artificielles comme auxiliaires de la pêche thonière en Océan Atlantique; Quel avenir? La Pêche Maritime 1291:655–659.

Fisheries Agency of Japan

1962–
1980a   Annual report of effort and catch statistics by area on Japanese tuna longline fishery. Research Division. Tokyo.

1962–
1980b   Annual report of effort and catch statistics by area on Japanese skipjack baitboat fishery. Research and Development Division. Tokyo.

Fisheries Research and Development Agency

1980–
1985   Annual report of catch and effort statistics and fishing grounds for the Korean tuna longline fishery 1975–1980. Pusan.

Floyd, J. M., and D. Pauly

1984   Smaller size tuna around the Philippines: Can fish aggregating devices be blamed? INFOFISH Marketing Digest 5:25–27.

Gillett, Robert

1985a   Traditional tuna fishing in Tokelau. Topic review no. 27. South Pacific Regional Environment Program. Noumea, New Caledonia.

1985b   Traditional tuna fishing at Satawal, central Caroline Islands. Mimeo.

Hallier, J. P.

1985   Purse seining on debris-associated schools in the western Indian Ocean. Report TWS/85/30. Food and Agriculture Organization. Rome.

Hilborn, R.

1985a   Spatial models of tuna dynamics in the western Pacific: Is international management necessary? Paper presented at the Second Workshop of Renewable Resource Management. December 9–12. Honolulu.

1985b   Interaction between surface and longline fisheries for yel-
             lowfin tuna. Internal report no. 7. Tuna and Billfish As-
             sessment Program. South Pacific Commission. Noumea,
             New Caledonia.

Ianelli, J.

1986   A method for estimating recruitment patterns of tunas to
             floating objects using removal data. Internal report no. 7.
             Tuna and Billfish Assessment Program. South Pacific
             Commission. Noumea, New Caledonia.

Lenarz, W. H., and J. R. Zweifel

1979   A theoretical examination of some aspects of the interac-
             tion between longline and surface fisheries for yellowfin
             tuna, *Thunnus albacares*. Fishery Bulletin 76:807–825.

Matsuda Y., and K. Ouchi

1984   Legal, political, and economic constraints on Japanese
             strategies for distant-water tuna and skipjack fisheries in
             Southeast Asian seas and the western central Pacific. In
             Memoirs of the Kagoshima University Research Center
             for the South Pacific 5(2). Kagoshima.

Sibert, J. R.

1984   A two-fishery tag attrition model for the analysis of mor-
             tality, recruitment, and fishery interaction. Internal report
             no. 7. Tuna and Billfish Assessment Program. South Pa-
             cific Commission. Noumea, New Caledonia.

South Pacific Commission Tuna Program

1986   A review of SPC's DWFN catch coverage in the SPC
             region. Working paper no. 19. Eighteenth Regional Tech-
             nical Meeting on Fisheries. South Pacific Commission.
             Noumea, New Caledonia.

Tuna Research Center

1973–   Annual catch statistics of Taiwan's tuna longline fishery
1983   1972–1982. Institute of Oceanography. National Taiwan
             University. Taipei.

PART II. DISTANT-WATER
TUNA FISHERIES

4.
Development of Japan's Tuna Fisheries

Norio Fujinami

INTRODUCTION

Japan's policies on tuna fisheries are part of its overall fisheries policy, which aims "to encourage optimum utilization of the resources for the development of the national economy."[1] This policy is pursued primarily by instituting resource management measures and establishing orderly marketing arrangements. Through time, Japan's fisheries policy has been modified to respond to the needs of the country's economy, to the needs of the fishing industry, and to the pressures of international events.

POLICY ASPECTS

Industrialization of Fisheries before 1945

Before World War II the Japanese economy depended mainly on primary industry (agriculture, forestry, and fisheries). More than half the population was employed in this sector, and farm and fish products were the staple foods. Indeed, most of the national intake of animal protein came from fish products because the people rejected meat and dairy products for religious reasons.

In the prewar period the chief roles of fisheries in the economy were to provide employment and to help maintain nutritional standards. The Japanese people frequently suffered from famine, especially in the north, where the harsh climate often decimated the harvest of rice and other crops. The government therefore adopted measures to expand fisheries production to maintain food supplies during shortages and to improve the distribution of food.

Japan's first fisheries administration arrangements were introduced by the central government in the 1870s. Following their introduction, a regal initiative was taken to organize fishermen and undertake experimental projects for improving the design and construction of fishing vessels. In the 1880s Japan's first fisheries training school was established, and offshore fishing operations were officially encouraged.

Japan enacted its first fisheries legislation in 1901 to control important fishing activities. The government also encouraged the mechanization of fishing vessels. At that time the Japanese tuna fleet consisted mostly of skipjack pole-and-line vessels. A few longliners, the product of technological advances in the late 1800s, were deployed as offshore fishing vessels. With governnment financial support, the first engine was installed on a skipjack pole-and-line vessel in 1903, the first refrigeration equipment in 1907, and the first radio in 1918.

Between 1900 and 1940 Japan's fisheries production increased sharply, from 440,000 tonnes in 1900 to 4.1 million tonnes in 1940.[2] Tuna production, however, merely doubled from about 100,000 tonnes to 200,000 tonnes because tuna fishing was confined to Japan's coastal and offshore waters and the country's demand for tuna was small.

Food Production Policies in the 1940s and 1950s

Between 1941 and 1945 Japan's offshore and distant-water fishing fleets were damaged by war. The fishing fleet declined from 350,000 vessels in 1941 to 280,000 small coastal craft in 1945. Partly because of the reduction in the fishing fleet and the loss of skilled manpower, Japan experienced severe food shortages throughout the 1940s and into the 1950s. The government therefore took measures to stimulate food production, including fisheries production. After 1945 the government also provided financial support for reconstructing port and shore facilities and made soft loans to fishermen for purchasing new vessels and gear.

Because of these initiatives, Japan's fisheries production recovered quickly after 1945; by 1950 the resources within the fishing areas authorized under the postwar occupation policy were fully exploited. When the peace treaty was signed in 1952, these fishing restrictions were lifted. Japan then launched its policy "from coast to offshore, from offshore to distant waters" to diversify its fishing activities so as to increase production to meet national demand. The policy succeeded. In 1955, just ten years after the war ended, production reached 4.9 million tonnes, exceeding the maximum production level of the prewar period. Tuna production also increased remarkably.

Industry Promotion in the 1960s

To meet Japan's increasing demand for tuna and to decrease fishing in fully exploited coastal and offshore waters, the government in 1952 initiated a policy of switching licenses from other fisheries to tuna fisheries. This policy was first applied to Japan's coastal purse seine and trawl fisheries; in 1960 it was extended to the salmon-trout driftnet fishery and in 1962 to salmon-trout driftnetters attached to motherships and coastal trawlers. The success of this policy showed that Japan's fisheries were interrelated and could therefore be managed under one general policy.

In 1960 Japan launched an epoch-making economic growth initiative premised on its high-quality but relatively low-cost manpower. The initiative achieved enormous success throughout the 1960s: the gross national product increased annually at a rate much higher than the world average. Japan's demand for tuna also increased sharply in the early 1960s, and the tuna fleets expanded their operations to meet it. By 1964 Japan's distant-water tuna fleets were operating all around the world (Figure 1). Furthermore, between 1960 and 1964 tuna production increased faster than overall fisheries production (Figure 2).[3]

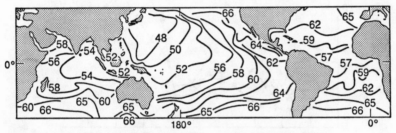

Figure 1. Expansion of tuna operation areas

Source: Federation of Japan Tuna Fisheries Cooperative Organizations.

Note: Figures are years from 1948 to 1966.

Through international expansion and the absorption of other fleets, the size of Japan's tuna fleets and of individual vessels increased. The number of licensed tuna vessels reached 3,000 in 1965.[4] At that time there were about 1,300 pole-and-line and longline vessels with no restrictions on distant-water operations, including 50 motherships ranging from 500 gross registered tonnes (GRT) to more than 9,000 GRT.[5] The enlargement and enhancement of Japan's tuna fleet were facilitated by technological advances in boatbuilding, fish detecting, vessel positioning, navigation, land-to-ship communication, and postharvest handling.

Despite improvements in the tuna fleet in the early 1960s, vessels could not meet the demand for sashimi because of problems in maintaining the quality of the fish on long fishing trips. Japan's longline fleet therefore concentrated on supplying foreign canning plants. In the 1960s there was strong international demand for canning material, and the Japanese government encouraged tuna exports to meet the demand and to generate foreign exchange. To rationalize their operations, the

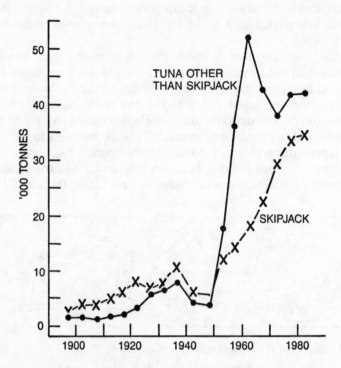

Figure 2. Production of tuna

Source: Fisheries production yearbook. Japanese Ministry of Agriculture, Forestry, and Fisheries.

Note: The graph shows four-year averages for 1894—97, 1908—11, 1922—25, and 1981—84. In other cases, five-year averages are used.

Japanese tuna fleets sought to establish base ports near their distant-water fishing grounds. Overseas bases were established in the Pacific islands (Federated States of Micronesia, Fiji, Palau, Papua New Guinea, and Solomon Islands) and elsewhere (Ghana, Madagascar, Malaysia, Maldives, Mauritius, Seychelles, and Spain-Las Palmas).

In the late 1960s, however, changes in Japan's economy and in the international tuna industry made overseas bases difficult to maintain. Vessel-operating costs were rising rapidly with the development of Japan's economy, fishing trips were being prolonged with the decline in daily catch rates, international demand for canning material was weakening, and demand for longline products for canning was declining (Figure 2).

Simultaneously, improvements in freezing technology enabled Japan's distant-water tuna fleets to switch to the production of sashimi-grade tuna rather than tuna for canning.[6] This breakthrough rendered foreign base ports largely redundant. Because sashimi landings had to be brought back to Japan for sale, it made financial sense to base the sashimi longline fleet in Japan. Moreover, the repeated handling of sashimi-grade tuna—aside from the costs of constructing, maintaining, and operating overseas transshipment facilities—reduced its market value. Hence most of the distant-water tuna vessels that had operated out of overseas bases opted to return to Japanese ports.

Having adjusted by relocating their operations, the Japanese fleets were then hit hard by the oil crisis in 1973.[7] To soften the blow, the government introduced special financial measures for the fishing industry and fostered the development of energy-efficient vessels, particularly for distant-water fishing operations.

Extended Jurisdiction in the 1970s and 1980s

The rising living standard stemming from Japan's rapid economic growth during the 1960s had the effect of decreasing the international competitiveness of some of its industries when growth slowed in the early 1970s. Difficulties were most pronounced for the labor-intensive primary industries—including fishing—and it was clear that structural changes were needed.

In 1977 coastal states around the world started to declare exclusive economic zones (EEZs). Japan proclaimed extended jurisdiction in 1977. The declaration of EEZs heralded a new era for Japan's distant-water tuna-fishing industry.

Prior to extended jurisdiction, Japan's distant-water tuna fleets had operated in the coastal waters of over 50 states in the Pacific and Indian Oceans, Southeast Asia, North America, Central and South America, Europe, and Africa. However, operations in most of these areas were relatively unimportant in comparison with tuna operations in the Pacific Ocean (Table 1).[8]

With the introduction of extended jurisdiction, Japan opted to establish relations with coastal states that had large, accessible tuna resources and to conclude access agreements with them to exploit their tuna resources. By 1986 it had made access agreements with 15 coastal states. The Japanese government negotiated agreements with the governments of Australia, Canada, France (overseas territories), Kiribati, the Marshall Islands, Morocco, New Zealand, Portugal, the Solomon Islands, South Africa, Tuvalu, and the United States. Japanese industry made agreements with three Pacific island governments: the Federated States of Micronesia, Palau, and Papua New Guinea.[9]

Toward the end of the 1970s the financial position of Japan's tuna fleets declined because of (1) drastically increased fuel costs, (2) rising

Table 1. Production by Japan's tuna fleets 1975–1982

	Pacific Ocean		Indian Ocean		Atlantic Ocean		Total	
	Tonnes	%	Tonnes	%	Tonnes	%	Tonnes	%
1975	470,000	86	673,000	90	574,000	87	592,000	86
1980	33,000	6	25,000	3	26,000	4	31,000	4
1981	43,000	8	54,000	7	60,000	9	70,000	10
1982	546,000	100	752,000	100	660,000	100	693,000	100

Source: FAO. Various years.
Note: Pacific Ocean—FAO Area Codes 61, 67, 71, 77, 81.
    Indian Ocean—FAO Area Codes 51, 57.
    Atlantic Ocean—FAO Area Codes 21, 27, 31, 34, 41, 47.

crew costs due to Japan's improved living standards, (3) increased vessel- and gear-construction costs, (4) extended duration of fishing trips due to declining catch rates, (5) restrictions imposed by extended jurisdiction and its associated costs (access fee payments, for example), and (6) stagnating market conditions for fisheries products in Japan due to changing food consumption patterns (Table 2).[10]

The changing conditions in the tuna industry required that Japan's tuna operations be rationalized and that distant-water fishing be undertaken by fewer but more efficient vessels. In consultation with the Japanese government, the industry in 1980 decided to reduce the number of distant-water tuna vessels by 20 percent and, by 1981–82, to introduce 20 high-seas purse seiners. The goals were achieved, and the composition of Japan's tuna fleet changed dramatically.

Table 2. Japan's animal protein intake 1955–1983

	Total		Fish/shellfish		Meat/eggs/ dairy products		Other	
	Amount[a]	%	Amount[a]	%	Amount[a]	%	Amount[a]	%
1955	16.9	100	13.0	77	3.4	20	0.5	3
1965	26.9	100	15.1	56	10.5	39	1.3	5
1975	35.7	100	17.5	49	17.6	49	0.6	2
1981	38.9	100	17.6	45	21.3	54	0.2	1
1982	38.7	100	17.1	44	21.6	55	0.2	1
1983	39.5	100	17.4	44	21.9	55	0.2	1

Source: Ministry of Agriculture, Forestry, and Fisheries 1985.
[a]In daily grams per capita.

The changes in the composition of Japan's tuna fleet are shown in Table 3. The number of distant-water pole-and-line and longline vessels with unrestricted overseas licenses declined by 26 percent, from 1,219 in 1970 to 900 in 1984. The number of pole-and-line vessels fell from 222 to 138 and the number of longliners from 997 to 762.

Table 3. Japan's licensed tuna vessels 1970–1984

	1970	1975	1980	1981	1982	1983	1984
Unrestricted distant-water[a]							
Pole-and-line	222	301	228	197	166	158	138
Longline	997	949	943	964	854	770	762
Total	1,219	1,250	1,171	1,161	1,020	928	900
Restricted offshore[b]							
Pole-and-line	497	438	343	342	377	374	276
Longline	1,092	808	757	752	605	592	633
Total	1,589	1,246	1,100	1,094	982	966	909
Purse seine vessels							
Single							
Pacific Ocean	3	10	13	23	32	32	32
Other	7	0	0	1	1	1	1
Group							
Pacific Ocean	0	0	4	7	7	7	7
Coastal[c]	100	100	100	100	100	100	100

Source: Ministry of Agriculture, Forestry, and Fisheries, Tokyo; Japan Far Seas Purse Seine Fisheries Associations and Federation of North Pacific Distant Purse Seine Fisheries Associations, Tokyo.

[a]Unrestricted overseas operational areas.
[b]Restricted areas of operations offshore.
[c]Approximate figures.

The reduction in the number of pole-and-line vessels and longliners with restricted offshore access was more pronounced than that of unrestricted distant-water fleets (Table 3). The number of pole-and-line vessels and longliners in the restricted offshore category declined from 1,589 in 1970 to 909 in 1984. The number of pole-and-line vessels fell from 497 in 1970 to 276 in 1984—a drop of 44 percent. The reduction in the number of longliners was 42 percent, from 1,092 vessels in 1970 to 633 in 1984.

The reduction in the number of tuna pole-and-line vessels for both distant-water and offshore operations produced a net loss in Japan's pole-and-line fishing capacity; that is, the total tonnage for this type of vessel declined. However, some of the capacity lost in the reduction of the longline fleet was regained by replacement vessels with larger capacities than the vessels being decommissioned.

Japan's single purse seine fleet developed rapidly over the 1970s (Table 3). Single seiners deployed in the Pacific Ocean—primarily in the waters of Micronesia and Papua New Guinea—increased from 3 vessels in 1970 to 32 vessels in 1984. The number of group seiners operating in the Pacific tuna fishery rose from four groups in 1980 to seven groups in 1984.[11] The number of coastal group seiners, however, remained constant at about 100 groups between 1970 and 1984.

The Japanese government supported the restructuring of the tuna fleet by undertaking research to improve vessel design (hull, engines, and refrigeration) and to automate pole-and-line and longline gears. This research proved fruitful: despite fleet reductions, tuna production did not fall (Table 4). In fact, it increased from 636,435 tonnes in 1975 to 863,510 tonnes in 1984—a rise of 36 percent. During the same period longline production increased by 10 percent, pole-and-line production by 9 percent, and purse seine production by 700 percent.

Table 4. Japan's tuna production 1975–1984 (in tonnes)

	Longline	Pole-and-line	Purse seine	Total
1975	291,832	322,582	22,025	636,439
1976	296,427	426,580	36,446	759,453
1977	319,508	343,665	40,952	704,125
1978	325,691	416,988	54,665	797,344
1979	326,518	361,459	60,841	748,818
1980	341,291	373,641	81,677	796,609
1981	313,323	294,109	95,945	703,377
1982	315,077	279,631	126,579	721,287
1983	327,376	277,572	164,176	769,124
1984	320,478	356,130	186,902	863,510

Source: Ministry of Agriculture, Forestry, and Fisheries 1985.

The financial position of Japan's tuna vessels, however, did not improve as a consequence of the rationalization program, largely because of the worldwide overproduction of tuna since the early 1980s. Except for certain sashimi species, tuna could lose its reputation and its value if overproduction continues. If it does, it will affect the commercial viability of Japan's tuna industry, and Japan will again have to reassess, and possibly reorganize, its tuna operations.

CURRENT INSTITUTIONAL ARRANGEMENT

To implement fisheries policies in Japan, the government has initiated (1) a licensing system, (2) the organization of fishermen, (3) an insurance program, (4) infrastructure improvement, and (5) the establishment and management of facilities for training, education, and research.

Licensing System

All major commercial fishing activities in Japan are regulated by the government through its licensing system. The system applies to all fishing ventures except small coastal vessels, fisheries in fixed locations (for example, oyster farms), and aquaculture. Longline and pole-and-line vessels greater than 20 GRT and purse seiners greater than 40 GRT engaged in harvesting tuna must be licensed. Licenses are renewed periodically, depending on prevailing resource and market conditions. The license compliance record of individual fishing vessels is also considered when licenses are reviewed for renewal.

Except for large longliners and pole-and-line vessels greater than 120 GRT, the government imposes restrictions on the areas where tuna vessels can operate (Figure 3) to ensure the safety of vessels, avoid concentration of fishing efforts, and facilitate orderly marketing procedures.

Although the tuna licensing system aims at achieving optimum use of resources and developing economically and commercially viable industries, attention has conventionally focused on maintaining viability by stabilizing the market.

Organization of Fishermen

The organization of fishermen is considered important in Japan for strengthening the social, economic, and political positions of individual fishermen and facilitating the implementation of government policy. Most fishermen are organized in a network of cooperative associations prescribed by the Fisheries Cooperative Association Law (1948). Fishermen not covered by this legislation are organized under other official or nonofficial arrangements.

Fishermen engaged exclusively in tuna fishing are organized into three groupings: fishermen with (1) longliners and pole-and-line vessels greater than 120 GRT, (2) longliners and pole-and-line vessels less than 120 GRT, and (3) single purse seiners. The first group covers about 200 pole-and-line vessels and 400 longliners, the second group 500 pole-and-line vessels and 300 longliners, and the third group 32 purse

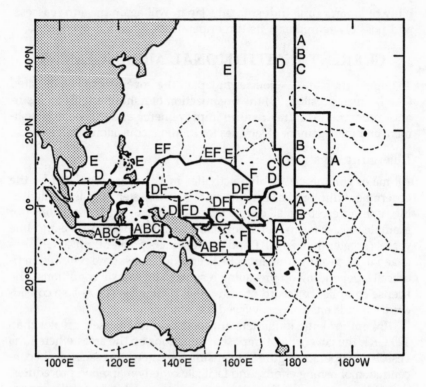

Figure 3.   Restrictions on operational areas of the Japanese tuna fleet

Notes: 1. Pole-and-line and longline vessels. Categories A, B, and C for vessels less than 120 GRT; categories D and E for vessels less than 80 GRT; category F for vessels less than 80 GRT based overseas. No restrictions apply to vessels greater than 120 GRT.

2. Purse seiners. Restricted to Pacific Ocean waters west of longitude 180° unless specifically authorized for experimental operations.

seiners. Group purse seiners are not included because they do not fish exclusively for tuna. They engage primarily in harvesting coastal species of fish (sardine and mackerel), harvesting tuna occasionally on a seasonal basis.

Insurance

The risks associated with fishing make it difficult for insurance companies to provide coverage at premiums fishermen can afford. The government therefore supports insurance programs for the fishing industry. Its support has contributed to the stable development and expansion of the fishing industry.

Because of the arduous nature of fishing operations, many industrial accidents occur each year. Of the 440,000 registered fishing vessels in Japan, approximately 240,000 vessels (55 percent) are covered by special government insurance programs. About 80,000 industrial accidents are reported under these programs annually; loss and damage to fishing vessels totals about $150 million per year.[12]

Infrastructure

The government invests large capital outlays for improving the fisheries infrastructure in Japan—harbors, shelters, shore facilities, shore protection arrangements, and so on. Two-thirds of the Fisheries Agency's annual budget of about $2 billion goes into the construction and maintenance of infrastructure. About 3,000 fishing harbors and shelters benefit from this expenditure. In addition, local governments and the infrastructure's principal beneficiaries—the fishermen—together contribute a like amount.

Training, Education, and Research

With few exceptions, Japan's institutions for fisheries training, education, and research are operated by central and local governments. Some 60 senior high schools and two colleges specialize in fisheries, and 15 universities have faculties of fisheries. These institutions do not provide specialized training in tuna fishery, but many of the senior high schools have training vessels of the types used for tuna fishing.

The central government has eight Regional Fisheries Research Laboratories. The Far Seas Fisheries Research Laboratory at Shimizu conducts tuna research; the Tohoku Regional Fisheries Research Laboratory at Shiogama specializes in skipjack research. Inshore fisheries research is carried out at experimental stations by 47 local governments. Some of these stations focus on tuna research.[13]

FUTURE POLICIES

Although Japan's tuna fisheries have experienced major difficulties, these have been largely overcome by the initiation of governmental policies with industry support. The problems now facing the tuna industry are more complex and international than earlier problems. For a long time all of Japan's fisheries production was directed to the domestic market. Foreign producers did not export fish to Japan because they did not have well-developed fishing industries, fish prices in Japan were low, and the Japanese demand for fisheries products was limited. The situation changed in the 1960s as Japan experienced rapid economic growth. Demand for fisheries products increased, and prices rose sharply. Recently, however, the Japanese market has tended to stagnate as a result of changing consumption patterns. In 1984 Japan's imports of fisheries products totaled $7 billion against its domestic fisheries production of $20 billion. The quantity and value of imported products—primarily high-priced products such as shrimp—have increased sharply in the past decade in line with Japan's rising income levels. Japan's longline catches and some pole-and-line products supply the country's sashimi market. In addition, about half the pole-and-line product and half the purse seine catch go to Japan's *katsuobushi* market—a market that differs from the international canned-tuna market.

Although the *katsuobushi* and canned-tuna markets differ, skipjack prices for the end products are interrelated. Since 1980 the international skipjack market has been soft because of rising production. The sashimi market, independent from other tuna markets, is confined to Japan and limited to about 350,000 tonnes of tuna per year. The chief exporters of sashimi-grade tuna to Japan are Korea and Taiwan; 90,000 to 120,000 tonnes are imported from these countries annually. Prices for sashimi tuna—especially red-meat tuna—drop sharply when inventories exceed 25,000 tonnes. Sashimi tuna production remains fairly constant because it requires special handling to meet the requirements of a discriminating market. However, the international supply of sashimi-grade tuna on the Japanese market is likely to increase, primarily because foreign producers are attracted by the market's high and stable prices.

The direction of Japan's tuna fishery—particularly of its distant-water fishery—will be affected by international events and growing competition. In order to remain competitive, Japanese fishermen will probably have to operate fewer but more efficient vessels. Japan will also have to intensify its research efforts in fish detection, fishing technology, postharvest handling, and vessel-operating efficiency in order to maintain or reduce production costs. Japan's fisheries administration system may also need to be overhauled. The system, which originated in the late 1800s to regulate the domestic fishing industry before Japan's international fisheries evolved, may now need to be modified to meet the challenge of changing circumstances.

NOTES

1. The views expressed in this paper do not necessarily represent those of the government of Japan. The paper is contributed in a personal capacity by the author.

2. The 1933 level of production was the maximum achieved in the prewar period.

3. Both domestic and international demand for tuna prompted the expansion of Japan's tuna fishing industry. Domestically there was strong demand for sashimi and *katsuobushi*; internationally there was strong demand for canning material.

4. This figure excluded small coastal tuna vessels under 20 GRT that were not regulated by Japan's licensing system.

5. These motherships are no longer operational.

6. Freezers were developed that could hold tuna at $-55°$ to $-60°$ C for the duration of fishing trips.

7. During 1974, the year after the first oil crisis, Japan experienced a negative growth rate in its gross national product.

8. Between 1975 and 1982 Japan's tuna production in the Pacific Ocean rose from 470,000 tonnes to 592,000 tonnes (Table 1). Pacific Ocean catches accounted for more than 86 percent of total annual tuna production. Indian Ocean production, while remaining relatively constant in volume between 1975 and 1982, trended proportionally downward. Atlantic Ocean production increased in both volume and proportion at the total.

9. Japan may increase the number of agreements it has with coastal states.

10. Japan's daily annual protein intake between 1955 and 1983 rose from 16.9 grams per capita to 39.5 grams per capita—an increase of 134 percent (Table 2). However, the fish and shellfish component of the total daily intake has remained relatively constant at around 17.5 grams per capita since 1975. The upward trend in total daily intake has been offset by increased intake of meat, eggs, and dairy products. In 1955 the daily intake of protein from this category was 3.4 grams per capita, but by 1983 it had reached 21.9 grams per capita. This trend to increased consumption of meat is expected to continue.

11. A group seiner usually consists of three vessels: a seiner of about 116 GRT, one or two carrier vessels of about 300 GRT, and an anchor vessel of about 30 GRT.

12. An exchange rate of ¥150=US$1.00 is assumed throughout the paper.

13. Tuna research is undertaken by stations in areas where tuna fishing is the main type of fishing.

REFERENCES

Food and Agriculture Organization (FAO)
  annual   Yearbook of fishery statistics. Various years. Rome.

Ministry of Agriculture, Forestry, and Fisheries
  1985   Indices of fishery statistics. Tokyo.

5.
Postwar Development and Expansion of Japan's Tuna Fishery

Yoshiaki Matsuda

INTRODUCTION

Long before World War II Japanese fishermen were fishing in Southeast Asian seas and in the western and central Pacific, developing most of the fishing grounds for tuna and skipjack tuna (*Katsuwonus pelamis*) in the region.[1] Their fishing did not conflict with local fishing because the resource was plentiful and local demand was small. By building fishing bases in the Pacific islands, the Japanese provided both technology transfer and jobs for islanders (Matsuda and Ouchi 1984).

However, with Japan's involvement in World War II the tuna and skipjack fisheries were put under the control of the military government in 1942. During the war Japan lost 60 percent of its tuna and skipjack vessels and all its overseas bases (Masuda 1963b). Many Japanese fishermen who had emigrated to Southeast Asia and the western and central Pacific returned to Japan after the war, leaving very few Japanese in the region.

Under the postwar occupation, production of food and of commodities for export became a top priority of MacArthur's policy. With the government promoting expansion of fishing grounds and subsidizing the construction of vessels, Japan's tuna and skipjack fisheries recovered quickly.

Although skipjack belongs to the tuna family, the Japanese have long treated it separately because it is produced, processed, and marketed differently from other tunas. Skipjack is conventionally caught by pole-and-line vessels, smoke-dried into *katsuobushi*, and marketed as a soup base or food flavoring; other tunas are usually caught by long-line vessels, kept fresh or frozen, and marketed as sashimi. For this reason, skipjack is discussed separately from other tunas in this paper.

71

LEGAL EXPANSION OF TUNA
AND SKIPJACK FISHING GROUNDS

Conventional Fisheries

Although all movements by Japanese fishing vessels were prohibited when World War II ended, wooden vessels were permitted to fish within 12 miles of Japan's coastlines by September 14, 1945 (Masuda 1963b). Thereafter the boundaries were extended four times: on September 27, 1945; on June 22, 1946; on September 19, 1949; and on May 11, 1950 (Figure 1). These boundary lines were collectively known as MacArthur

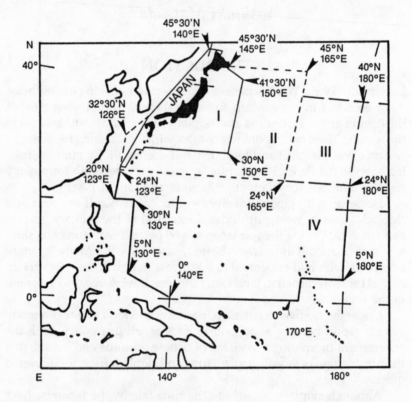

Figure 1. The MacArthur Lines: Legal expansions of Japanese fishing
          zones 1945–1952
    I. First expansion, September 27, 1945
   II. Second expansion, June 22, 1946
  III. Third expansion, September 19, 1949
   IV. Expansion for mothership operations, May 11, 1950

Source: Adapted from Masuda 1963b.

Lines. All restrictions on the movements of fishing vessels were lifted on April 25, 1952.

After the 1952 lifting of the last MacArthur Line, Southeast Asian seas and the Pacific were divided according to terms of the Japanese 1949 licensing law into A, B, and C zones for tuna longline and skipjack pole-and-line fisheries (Figure 2). Vessels were designated to fish in the three zones as follows:

Zone A: medium-scale vessels (20–100 GRT)

Zone B: distant-water vessels (above 100 GRT)

Zone C: mothership operations

Vessels under 20 GRT were free to fish anywhere.

It is important to understand that the boundary line between zones A and B is merely the outer limit of safety for smaller vessels registered

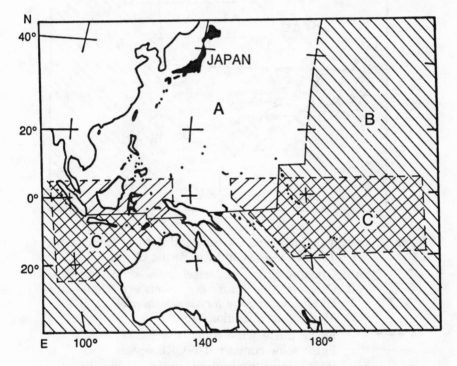

Figure 2. Legally defined Japanese tuna and skipjack fishing zones after the lifting of the MacArthur Line in 1952

A. Medium-scale tuna and skipjack fishing zone

B. Distant-water tuna and skipjack fishing zone

C. Mothership operations zone

Source: Adapted from Masuda 1983a.

for offshore fishing. Vessels registered for distant-water fishing may fish in both zones; however, they fish chiefly in zone B.

On July 10, 1953, in an effort to promote distant-water tuna and skipjack fisheries, the government enacted legislation permitting vessels over 70 GRT to fish in zone B. The government also encouraged the construction of larger fishing vessels (Masuda 1963b).

Regulations for the tuna and skipjack fisheries have been revised many times, but the designation of zones A and B has remained constant (Figure 3). Some of the revisions have changed the size designa-

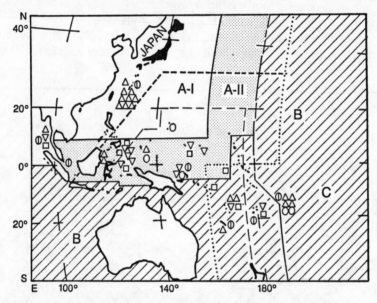

Figure 3. Legally defined Japanese tuna and skipjack fishing zones in Southeast Asian seas and the Pacific Ocean 1980

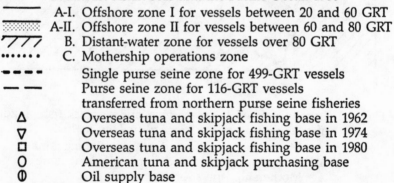

A-I. Offshore zone I for vessels between 20 and 60 GRT
A-II. Offshore zone II for vessels between 60 and 80 GRT
B. Distant-water zone for vessels over 80 GRT
C. Mothership operations zone
Single purse seine zone for 499-GRT vessels
Purse seine zone for 116-GRT vessels transferred from northern purse seine fisheries
△ Overseas tuna and skipjack fishing base in 1962
▽ Overseas tuna and skipjack fishing base in 1974
□ Overseas tuna and skipjack fishing base in 1980
○ American tuna and skipjack purchasing base
⦿ Oil supply base

Source: Adapted from Masuda 1963a and Suisan Sha 1972, 1973, 1974a, 1974b, and 1981.

tions for vessels in each zone, in effect redefining "medium-scale vessels" and "distant-water vessels." A 1967 revision divided zone A in two, zone A–I for smaller vessels and zone A–II for larger ones. A revision in 1977 expanded zone A–II in the Banda Sea by agreement between Indonesia and Japan.

A new set of laws governing offshore tuna and skipjack fishing grounds was adopted at the fourth renewal of licenses in 1981 (Figure 4). Offshore fishing grounds were expanded to the east and the south, and zone A–I–3 was established for Okinawa fishermen. Entry to zone A–II–2 (east of 180°) was limited to 90 vessels.

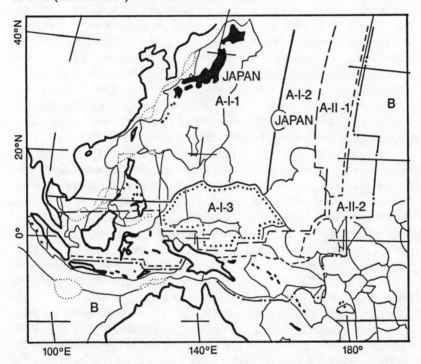

Figure 4. Legally defined Japanese tuna and skipjack fishing zones 1981

———— Zone A-I-1 for vessels between 20 and 80 GRT
— — Zone A-I-2 for vessels between 20 and 80 GRT
········ Zone A-I-3 for Okinawan vessels between 20 and 80 GRT
– – – – Zone A-II-1 for vessels between 80 and 120 GRT
—·—·— Zone B for distant-water fishing
———— Exclusive economic zone (200-mile zone)
············ Disputed area

Source: Adapted from Suisan Shincho Sha 1983

The offshore tuna and skipjack fisheries in zone A originated in the 1949 free-fishing right granted to vessels under 20 GRT. The distant-water tuna and skipjack fisheries in zone B developed from the original medium-scale tuna longline and skipjack pole-and-line fisheries.

In addition to its home-based tuna and skipjack fisheries, Japan has three other kinds of tuna and skipjack fisheries: a tuna longline fishery with motherships, foreign-based fisheries, and purse seine fisheries.

Other Operations

Mothership operations got special attention in 1950 under the occupation (Figure 1). In fact, the fourth expansion of fishing grounds was just for this fishery. After the lifting of the MacArthur Line, mothership fishing grounds were divided into two areas—one in Southeast Asian seas, the other in the South Pacific (Figure 2). The area east of 170° in the South Pacific was opened to mothership operations in 1956, the area east of 170° in the North Pacific in 1957 (Masuda 1963a). The use of motherships was characteristic of postwar tuna fishing. Zone C eventually became mothership fishing grounds (Figure 3).

Foreign-based tuna and skipjack fisheries included joint ventures, chartered vessels, sales contracts, and technical assistance arrangements. Most of these are private enterprises established under Japan's general business laws. To promote foreign-based tuna and skipjack fisheries, such semigovernmental organizations as the Overseas Fisheries Cooperation Foundation (OFCF) and the Overseas Cooperation Foundation (OCF) have provided technical and economic assistance. OFCF and OCF loans cover 70 percent of the capital investment of joint ventures.

In the late 1970s purse seining gained popularity in the Japanese tuna and skipjack fisheries. Their legal areas of operation in 1980 are shown in Figure 3. The number of 499-GRT Japanese single purse seiners in the western central Pacific increased from 13 in 1980 to 33 in 1983. In addition, seven fleets, each consisting of 116-GRT purse seiners with two transportation vessels and a search vessel, transferred from the northern fishery and now fish for four months beginning in April each year.

Tuna and Skipjack Fisheries

After the war Japanese tuna and skipjack fisheries developed quickly within the legislated limits on fishing grounds and restrictions on entry. With low-interest loans from the government, fishermen often enlarged their vessels, and by 1950 the fishing grounds were being extensively used (Figure 5). As of May 1958, 1,104 Japanese conventional tuna and skipjack fishing vessels over 20 GRT were in opera-

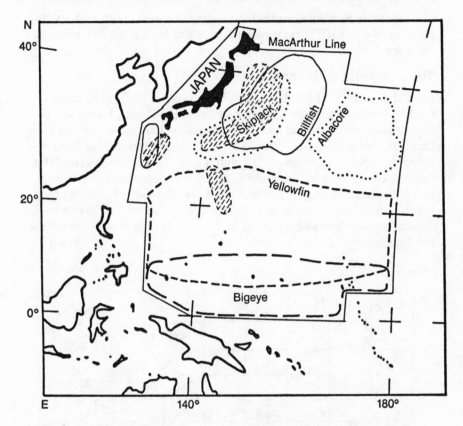

Figure 5. Major Japanese tuna and skipjack fishing grounds 1950

Source: Adapted from Nikkatsuren 1962

tion. Their number increased to 2,975 in 1967, then decreased to 2,271 in 1985 (Japan Fisheries Agency 1968 and 1986).

Although Japan expanded its fishing operations all over the world in the 1960s, its tuna and skipjack fisheries were hit hard by the rises in oil prices in 1973 and 1978. The necessity for containing costs by reducing oil consumption made energy-saving devices and less-distant fishing grounds central to Japanese fishing. As a result the Pacific region has become more important to tuna longline fishermen than ever.

In addition, the United Nations Law of the Sea Convention led to a worldwide boom in claims to extended maritime jurisdiction. Many countries in and surrounding the Pacific Ocean claimed their 200–mile exclusive economic zones or fishing zones (EEZs or EFZs), and Japanese fishermen could no longer fish their waters without paying fees.

Consequently they focused their operations on areas with high catch-per-unit-of-effort (CPUE) ratios and/or fish of high economic value in the Pacific.

Tuna Longline Fishery

After the removal of the last MacArthur Line in 1952, tuna longline fishermen extended their operations all over the Pacific, taking advantage of their mobility and low fuel prices. By 1964 most parts of the Pacific between 40° north latitude and 40° south latitude had been explored by Japanese longline fishermen (Figure 6). They moved into the Atlantic Ocean, the Indian Ocean, and the waters south of Australia and New Zealand during the late 1960s and the early 1970s. As they spread out, their activity in the Pacific lessened, and Korean and Taiwanese tuna longliners moved in to develop their own fisheries in the Pacific.

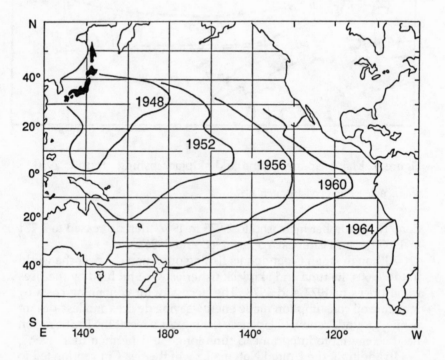

Figure 6. Expansions of Japanese tuna longline fishing in the Pacific 1948–1964

Source: Adapted from Suzuki et al. 1978

The oil crises in the 1970s drove many Japanese fishermen back to the Pacific fishery. Figure 7 shows the area of the yellowfin tuna longline fishery from April to July, the most productive quarter, in 1980. During this period the highest CPUE was 4.5 percent; the average was only 1.2 percent.

Skipjack Pole-and-Line Fishery

Immediately after the war the skipjack fishery developed rapidly because the operation was labor-intensive and did not require sophisticated equipment. However, because this fishery requires live bait, skipjack fishing was limited to grounds in the western North Pacific before 1970. Even now it does not go beyond the western central Pacific, in spite of major improvements in live-bait transportation (Figure 8).

Tuna Longline Fishery with Mothership Operations

Taiyo Gyogyo Company conducted a tuna-fishing feasibility study with a mothership, the *Banshu Maru* (1,066 GRT) in 1948, and Nihon Suisan

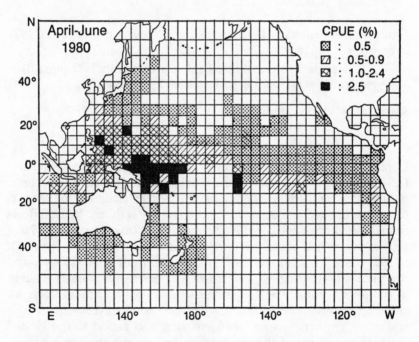

Figure 7. Japanese yellowfin longline catch per unit of effort in the Pacific 1980

Source: Adapted from Japan Fisheries Agency 1982.

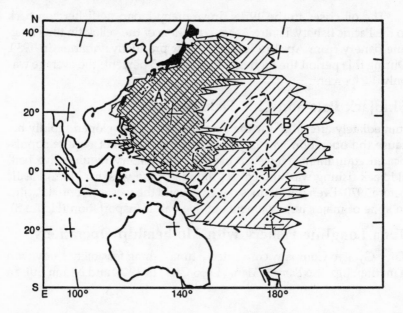

Figure 8. Expansions of Japanese skipjack fishing in the Pacific 1979

𝒮𝒮𝒮𝒮 Reported pole-and-line fishing grounds in 1970 (area A)
⁄⁄⁄⁄ Reported pole-and-line fishing grounds in 1977 (area B)
– – – Purse seine fishing grounds in 1979 (area C)

Sources: Japan Fisheries Agency 1972, 1979; Suisan Sha 1980.

Company conducted a similar study with the mothership *Kaiko Maru* (2,940 GRT) in 1949. Although the studies did not prove the economic feasibility of mothership operations, the 1950 extension of the MacArthur Line opened the ocean to mothership operations. In 1950 fleets from three companies—Taiyo Gyogyo Company, Hoko Suisan, and Nansei Suisan—fished about 6,500 tons of tuna from zone IV (Figure 1). In 1951 three fleets from Nansei Suisan, two from Taiyo Gyogyo, and one from Nihon Suisan—a total of 96 associated fishing vessels—caught 11,000 tons of tuna. The Nansei Suisan and Nihon Suisan fleets included eight small catcher boats, which carried fresh yellowfin to their respective motherships for quick freezing and export to the United States. The success of this venture opened export markets for tuna.

This fishery prospered in the early 1960s but terminated in 1966 because the North Pacific fisheries offered higher economic incentives for mothership operations than the tuna fishery did (Masuda 1963a).

Foreign-Based Fisheries

The first foreign-based tuna and skipjack fisheries after the war were established by Mitsubishi Shoji Company in 1952. The company purchased tuna and skipjack from Japanese fishermen and sold them to Star-Kist in American Samoa. By 1962 there were 38 such foreign-based tuna and skipjack fishing operations, including nine in Okinawa, one in Malaysia, two in Sabah, four in American Samoa, and two in Vanuatu (New Hebrides) (Figure 3).

An additional 14 bases were established by 1974—among them three in Papua New Guinea, three in Indonesia, and one each in Thailand, Truk, Solomon Islands, Pohnpei, Palau, and the Philippines.

Oil supply bases were established in Penang, Malaysia; Singapore; Darwin, Fremantle, Hobart, and Sydney in Australia; Wellington, Auckland, and Littleton in New Zealand; Noumea in New Caledonia; Suva in Fiji; Rabaul in Papua New Guinea; Papeete in Tahiti; and Honolulu in the United States by 1974 (Suisan Sha 1974b). Japanese fishermen have also used American tuna and skipjack purchasing bases at Pago Pago in American Samoa, Rabaul in Papua New Guinea, Koror in Palau, and Guam and Honolulu in the United States.

Purse Seine Fishery

Purse seining is considered more economical than conventional tuna and skipjack fishing methods because it does not require bait, is more labor-efficient, and consumes less fuel. However, Japanese tuna and skipjack fishermen resisted purse seining because of lack of experience, competition with established purse seine operations, inadequate technology for deep-water purse seining, high construction costs for vessels, and fear of depleting the resource.

Purse seine test operations in the western central Pacific began around 1960. After Taiyo Gyogyo's *Taikei Maru* (210 GRT) made a successful trip in 1964, three 300-GRT purse seiners joined the operation. However, its economic feasibility was not proven (Suisan Sha 1968). In 1967 the Japanese government conducted an investigation of potential fishing grounds for tuna and skipjack purse seiners. The Japan Marine Resource Research Center, established in 1971, has been conducting economic feasibility studies since 1974 of year-round purse seining for tuna and skipjack in the western Pacific. The number of purse seiners increased to 13 by 1980.

Some of the technological problems have now been solved, and the economic feasibility of purse seining has been confirmed. In connection with the vessel-reduction plan led by the Japanese Federation of Tuna and Skipjack Fisheries Cooperatives (Nikkatsuren), conventional fishermen built ten 499-GRT purse seiners in the early 1980s.

Conventional purse seine fishermen based on 110-GRT vessels in northeastern Japan expanded into purse seining for skipjack after good experimental results in 1980. Although Taiyo and Kyokuyo were the ones who established the Japanese tuna and skipjack purse seine fishery, other large fishing companies are also seeking investment opportunities in tuna and skipjack purse seiners.

INDUSTRIAL TRENDS

Statistical Summary

After the war, Japanese tuna and skipjack production increased rapidly (Figure 9). The 1951 tuna and skipjack catch of 235,912 tonnes exceeded the highest prewar catch of 202,439 tonnes in 1940 (Matsuda and Ouchi 1984). With the rapid expansion of the tuna longline fishery, total tuna and skipjack production reached 722,364 tonnes in 1962. However, tuna production has decreased since then, while the skipjack catch has increased. As a result, total production has leveled off.

The rise and fall of different fishing methods in the Southeast Asian seas and the Pacific Ocean are shown in Table 1. Right after the war, Japan-based pole-and-line fishing was most popular. But this produc-

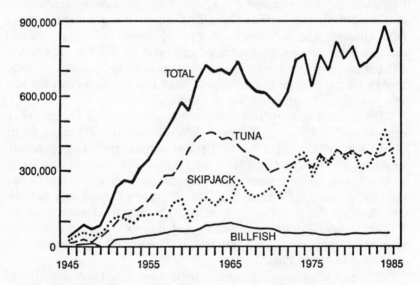

Figure 9. Japanese tuna and skipjack production 1945–1985 (in tonnes)

Source: Adapted from Japan Ministry of Agriculture, Forestry, and
Fisheries Annual and 1979.

Table 1. Japanese tuna and skipjack production in Southeast Asian seas and the Pacific by type 1951–1985 (tonnes)

	Japan-based tuna longline fishing[a]		Japan-based skipjack pole-and-line fishing		Tuna longline fishing with motherships		Foreign-based fishing[b]		Purse seine fishing	
	Tuna[c]	Skipjack	Tuna	Skipjack	Tuna[c]	Skipjack	Tuna[c]	Skipjack	Tuna[c]	Skipjack
1951	90,199	203	27,135	96,214	9,915	18	0	0	375	289
1955	175,842	536	35,207	92,524	11,699	28	7,245	0	13,163	2,363
1960	321,507	1,267	32,193	70,428	21,716	0	14,909	0	11,152	3,620
1965	309,463	1,092	51,698	127,436	4,251	0	31,473[d]	35[d]	14,208	4,637
1970	302,608	1,544	34,390	187,438	0	0	8,744[d]	27[d]	8,582	7,081
1975	260,484	224	67,264	244,348	0	0	0	15,379[e]	13,490	8,504
1980	301,904	82	65,209	295,568	0	0	0	46,531[e]	32,976	49,982
1981	281,555	66	44,778	237,685	0	0	0	14,098[f]	52,051	43,904
1982	287,190	56	48,597	218,146	0	0	0	8,661[g]	54,392	73,154
1983	295,475	1,196	40,757	224,179	0	0	0	14,633[g]	48,959	114,980
1984	285,988	108	45,955	297,591	0	0	0	15,523[g]	55,797	134,641
1985	304,347	217	40,184	181,821	0	0	0	n.a.	64,656	125,279

Sources: Compiled from Irabu-mura 1981; Japan Ministry of Agriculture, Forestry, and Fisheries 1977–1986 and 1979; Kataoka and Matsuda 1983; and Solomon Islands Government 1985.

[a]Includes Indian Ocean operations.
[b]Includes American Samoa, Vanuatu, and Noumea bases only.
[c]Includes billfish.
[d]Includes landings from bases along Indian Ocean coasts.
[e]Includes only Irabu-mura (Okinawa) landings.
[f]Includes only Okinawa fishermen's landings to Solomon-Taiyo Company.
[g]Includes only half of total landings in Solomon Islands.

tion was soon taken over by Japan-based tuna longline fishing. Tuna production exceeded skipjack production in 1952. The gap widened until 1965, when tuna production fell off because of lower productivity and skipjack production surged ahead because of improvement in live-bait transportation and development of overseas bases. In 1980 production of tuna and skipjack were almost equal at about 380,000 tonnes. The rapid growth in production corresponded to the rapid expansion of tuna longline fishing during the 1950s and of skipjack pole-and-line fishing during the early 1970s.

Tuna longline fishing with motherships peaked in 1962, then ended in 1966 when it lost its economic viability. Although foreign-based fishing at first centered on tuna, it shifted to skipjack in the 1970s. However, foreign-based skipjack production has declined rapidly during the 1980s.

After a long trial-and-error period, production from purse seine fishing has grown to over a third of total skipjack production.

Fishing Activities in Southeast Asian Seas

Statistics are not available for tuna longline operations before 1962 and for skipjack pole-and-line operations before 1968. However, Japanese were fishing for tuna and skipjack in Southeast Asian seas before 1962 (Masuda 1963a).

After the removal of the final MacArthur Line in 1952, the *Kaiko Maru* (Nihon Suisan, 2,940 GRT) with ten longliners fished in Celebes, Moluccas, and the Banda Sea. The *Ginyo Maru* (Hokkaido Kosha, 3,840 GRT) with ten small catcher boats under 20 GRT and the *Saipan Maru* (Taiyo Reito, 3,738 GRT) with five longliners fished south of Java, and the *Tenyo Maru* (Taiyo Gyogyo, 3,689 GRT) with 16 catcher boats fished in the Andaman Sea and waters off the Nicobar islands in 1954. Kushi-kino tuna longliners and Makurazaki skipjack pole-and-line vessels were active at the same time. The South China Sea was also regarded as good fishing grounds for large tuna, especially from September to May. The average CPUE in the South China Sea varied from 1.5 to 4.5 percent by month (except for June to August) from 1930 to 1956, excluding the years 1942 to 1950.

Available statistics show that there were continuous tuna longline fishing operations in Southeast Asian seas from 1962 on and skipjack pole-and-line operations from 1968 to at least until 1977 (Matsuda and Ouchi 1984).

The Japanese consider most Southeast Asian waters as offshore, not distant-water, fishing grounds because the fishing is seasonal and the vessels used are under 80 GRT (Figures 2–4). The major species caught by longline were yellowfin and bigeye tuna, with annual production fluctuating between 17,000 and 21,000 tons; skipjack production varied from 1,000 to 6,000 tons (Matsuda and Ouchi 1984).

Fishing Activities in the Pacific

The western central Pacific Ocean has been a major fishing ground for Japanese tuna and skipjack fishermen. Tuna longline fishing with motherships commenced in 1948; it expanded into area IV (Figure 1) as early as 1950 and 1951. By 1961 the industry had grown to five fleets with 209 associated tuna longliners and ice supply vessels, which fished for 522 fleet-days between May and March, catching 28,933 tonnes of fish (including 26,050 tonnes of tuna) in the South Pacific. However, mothership operations declined gradually with the increase in larger longliners with quick freezers on board and the concomitant decrease in small longliners to join the fleets and with the development of foreign-based tuna and skipjack fisheries.

By the end of 1952 Japan-based tuna longliners had expanded their fishing grounds in the western central Pacific. The development of tuna fishing based in American Samoa was a by-product of such activities. Records from 1942 to 1974 (Figure 3) show that Japan habitually carried on foreign-based tuna fishing from American Samoa, Santo Island (Vanuatu), Noumea, and Fiji, while its tuna longline operations in the Pacific were continuous. Estimates of annual catches of tuna by Japanese longliners within the 200-mile EEZs of member countries of the South Pacific Commission varied from 29,763 tonnes to 42,406 tonnes during the period 1972–1976 (Klawe 1978). The major fishing grounds were Micronesia and the waters of Kiribati, Papua New Guinea, Solomon Islands, Nauru, and French Polynesia.

Makurazaki pole-and-line vessels fished in waters off Okinawa, Taiwan, and the Philippines soon after the removal of the MacArthur Line. However, southward expansion by Japanese skipjack pole-and-line vessels was long limited to the southwestern North Pacific; expansion to the South Pacific came only after 1970. That expansion was rather rapid when it did begin: the skipjack catch from the South Pacific Commission area jumped from 59,112 tonnes in 1972 to 134,591 tonnes in 1974 (Kearney 1979). The major fishing grounds for pole-and-line vessels, as for longliners, were Micronesia and the waters of Kiribati, Papua New Guinea, Solomon Islands, Nauru, and Tuvalu.

Under access-fee arrangements, Japanese fishermen are now fishing in the waters of the United States, the Republic of Belau, Federated States of Micronesia, Marshall Islands, Kiribati, Solomon Islands, Papua New Guinea, France, Australia, and New Zealand.

Limiting Factors

The postwar expansion of Japanese tuna and skipjack fisheries has not been smooth. On March 1, 1954, the *Fukuryu Maru V*, a tuna longliner, encountered radioactive fallout from a U.S. hydrogen bomb test over Bikini Atoll in the Marshall islands. The mounting number of polluted tuna catches landed in Japanese ports following the test caused a panic in the tuna market (Suisan Shuho Sha 1954). In less than a month, tuna prices in Tokyo dropped 21.9 percent for offshore tuna, 29.4 percent for yellowfin, and 49.4 percent for bigeye. All 425 polluted vessels were subsequently identified, and 326.3 tonnes of tuna was discarded. Polluted tuna was reported in the eastern seas of the Philippines and Taiwan, more than 1,000 miles from Bikini. Although none was landed from this region, tuna exports suffered from an increase in canceled contracts.

The 1960s were relatively smooth years for the tuna and skipjack fisheries, and the industry devoted itself to exploiting fishing grounds all over the world by establishing an extensive infrastructure. However, the tuna longline fishery had a very low CPUE during the 1970s (Japan Fisheries Agency 1982).

Other problems have included general inflation and cost inflation, particularly for labor and fuel; difficulties in exporting to the United States because of dollar devaluation in 1972 and changes in Food and Drug Administration (FDA) regulations regarding the mercury content in tuna; growing competition from Taiwanese and Korean tuna fleets in fishing and marketing; a depressed domestic economy caused by oil crises in 1973 and 1978; mounting concern for the environment; and the advent of 200-mile exclusive zones. The consequence has been a decline in profits since 1974 (Matsuda and Ouchi 1984).

The economic impact of 200-mile zoning on Japanese tuna and skipjack fisheries has been immense. In 1977, 48 percent of Japan's tuna catches and 41 percent of its skipjack catches came from within 200 nautical miles of the coasts of 54 foreign countries. The declaration of 200-mile zones depressed the industry during the years that followed. Government and private efforts—including price supports, attempts to reduce the number of licenses, and 30-day moratoriums on skipjack fishing—have had limited success. The price of skipjack recovered in late 1979 and early 1980 but soon dropped again because of overproduction, export problems, and increasing landings by purse seiners.

The demands placed on Japanese fishing fleets by coastal states are increasing continually under the 200-mile regime. There are royalty or entry fees, registration fees, quotas, excess-catch fees, less favorable terms for joint-venture enterprises, requests to expand imports of agricultural products to Japan, and requests for development assistance from Japan. Japan has responded by reorganizing the Ministry of Agriculture and Forestry into a new Ministry of Agriculture, Forestry, and Fisheries in 1978 and restructuring the government fisheries administration.

The chief participants in Japanese tuna and skipjack fisheries are individual fishermen, small- or medium-sized companies, and members of the Japanese Federations of Fisheries Cooperatives (Zengyoren), the Tuna and Skipjack Fisheries Cooperatives (Nikkatsuren), and the Offshore Tuna and Skipjack Fishermen's Association (Kinkatsukyo). Most of them own one or two vessels in the 50- to 500-GRT range. In 1978, 651 enterprises were involved in distant-water tuna and skipjack fishing and 994 enterprises practiced offshore fishing. Of these, 77 fished in both fishing grounds (Suisan Sha 1979). The large fishing companies such as Taiyo, Hokoku, and Kyokuyo supplied only 25,915 tons in 1977, or 3.7 percent of the total Japanese catch.

The tuna and skipjack fisheries consumed about 22 percent of all oil used by the Japanese fishing industry in 1974. The oil crises wiped out their profits in spite of their efforts to use energy-saving devices.

The economic and political events of the 1970s seriously weakened small Japanese fishing enterprises. They could not accumulate enough capital for joint ventures, so they had no choice but to continue fishing for survival. They preferred fee fishing over joint ventures, not only because the government helped bear the burden but also because they found it hard to deal with joint ventures. Thus the 20-percent vessel-reduction plans proposed by Nikkatsuren, Zengyoren, and Kinkatsukyo in the early 1980s were inevitable.

Although optimism prevails about the potential of skipjack resources in the world ocean, fishing effort no longer correlates positively with harvest in the western central Pacific (Suisan Shincho Sha 1985). Large tuna are already being caught at the maximum sustainable yield. Most tuna and skipjack joint ventures failed in the late 1970s and early 1980s. In 1982 both Kyokuyo and Kaigai Gyogyo withdrew from their ventures in Papua New Guinea because of new taxes on corporations (Kataoka 1984). Itochu Shoji withdrew from the Pacific Fishing Company (PAFCO), one of the most successful companies in the South Pacific, in 1986. Only Taiyo Gyogyo made a second ten-year contract with Solomon Islands in 1982. This Japanese company is the only active fishing company in the Pacific islands. Stagnant market prices for tuna and skipjack and increasing local demands on businesses are critical causes of such failures.

Thus the limiting factors are many, and legal and political constraints tend to aggravate the economic constraints.

FUTURE IMPLICATIONS

The Japanese tuna and skipjack fishing industry has taken drastic steps since 1974 to cope with the problems of cost inflation, stagnant prices, and extended maritime jurisdiction. A number of countermeasures have strengthened the flexibility of the industry and softened the impacts of the problems. These measures include using energy-saving operations and devices, using mechanisms for stabilizing demand and price, withdrawing fishing vessels, increasing participation in purse seine fisheries, and adjusting offshore fishing grounds.

Bilateral arrangements that link fee fishing, overseas technical and economic cooperation, and joint ventures have been employed as external strategies. However, arrangements are becoming more and more difficult because of coastal nations' desires to raise fishing fees and to have a stake in joint ventures.

Although the Japanese government is trying to ease the transition for tuna and skipjack fishermen, it is also implementing a multisectoral policy by assisting joint ventures and allowing purse seiners to join the tuna and skipjack fleets, thereby creating new tuna and skipjack fishermen at a time when the situation has worsened for conventional tuna and skipjack fishermen.

Extended maritime jurisdiction poses problems not only for the Japanese but also for coastal nations themselves. Extended jurisdiction benefits coastal nations, but it also burdens them because no single county has enough resources to manage the tuna and skipjack stocks within its EEZ.

Joint arrangements offer advantages over going it alone. However, each participating party must understand the interests, capabilities, and limitations of the other. Goals must be reasonable, and growth must proceed step by step in a way that benefits both sides.

Because tuna and skipjack are highly migratory, the time has come to consider them part of the common heritage of mankind. It is essential to face up to the conflicts between competing interest groups. This is a time to learn from each other, not to take advantage of each other. Past experience must be reviewed critically, and a new international order of cooperation for sound management of tuna and skipjack resources must emerge.

NOTE

1. This research is based on a study undertaken at the East-West Center's Environment and Policy Institute. Assistance was also received from faculty members in Kagoshima University's Faculty of Fisheries and its Research Center for the South Pacific.

REFERENCES

Irabu-mura

1981   Status of Irabu-mura fisheries. 6 pp.

Japan Fisheries Agency

1968   Statistical tables of fishing vessels. General report no. 19.

1986   Statistical tables of fishing vessels. General report no. 38.

1972   Annual report of effort and catch statistics by area: Japanese skipjack bait-boat fishery 1970.

1979   Annual report of effort and catch statistics by area: Japanese skipjack bait-boat fishery 1977.

1982   Annual report of effort and catch statistics by area: Japanese tuna longline fishery 1980.

Japan Ministry of Agriculture, Forestry, and Fisheries

Annual   Annual report on fisheries and aquaculture production statistics 1976–1985.

1979   Cumulative fisheries statistics 1894–1975. Vol. 2:2–5.

Kataoka, Chikashi

1984   Localization policy and development of industrial fisheries in Papua New Guinea. In Memoirs of the Kagoshima University Research Center for the South Pacific 5(1): 66–85. Kagoshima.

Kataoka, Chikashi, and Yoshiaki Matsuda

1983   Tuna and skipjack fisheries development in the South Pacific: Case of Fiji and Solomon Islands. Gyogyo Keizai Kenkyu 28 (3):46–60.

Kearney, Robert E.

1979   An overview of recent changes in the fisheries for highly migratory species in the western Pacific Ocean and projections for future development. South Pacific Bureau for Economic Cooperation no. 17. Suva, Fiji.

Klawe, William L.

1978   Estimates of catches of tunas and billfishes by Japanese, Korean, and Taiwanese longliners from within the 200-mile economic zones of the member countries of the South Pacific Commission. SPC occasional paper no. 10. Noumea, New Caledonia.

Masuda, Shoichi

1963a   Status of fisheries production. In Tuna and skipjack fisheries in Japan. Edited by Shoichi Masuda. The Japanese Federation of Tuna and Skipjack Fisheries Cooperatives (Nikkatsuren) and the Japanese Tuna and Skipjack Fishermen's Association (Kinkatsukyo). pp. 343–504.

1963b   Fisheries policy. In Tuna and skipjack fisheries in Japan. Edited by Shoichi Masuda. pp. 755–779.

1975   Tuna and skipjack fisheries under the 200-mile regime. Suisan Shuho 796:14–19.

Matsuda, Yoshiaki, and Kazuomi Ouchi

1984   Legal, political, and economic constraints on Japanese strategies for distant-water tuna and skipjack fisheries in Southeast Asian seas and the western central Pacific. In Memoirs of the Kagoshima University Research Center for the South Pacific 5(2):151–232. Kagoshima.

Nikkatsuren

1962   Status of tuna and skipjack fisheries. Skipjack and tuna: 148–158.

Solomon Islands Government

1985   Fisheries development annual report 1984. 85 pp.

Suisan Sha

1968   Purse seine fishery. Fisheries Yearbook 1968:148–153.

1972   Capital integration in fisheries. Suisan Shuho 660:23–24.

1973   Tuna and skipjack fisheries. Fisheries Yearbook 1973:56–76.

1974a   A table of Japanese fisheries joint ventures. Suisan Shuho 726:104–125.

1974b   Tuna and skipjack fisheries. Fisheries Yearbook 1974:67–77.

1979   Tuna and skipjack fisheries. Fisheries Yearbook 1979: 126–130.

1980   Purse seine fishery. Fisheries Yearbook 1980:138–139.

1981   A table of Japanese fisheries joint ventures. Fisheries Yearbook 1981:143–148.

Suisan Shincho Sha

1983   Operation map for offshore tuna and skipjack fisheries: Essentials for tuna and skipjack fishermen. pp. 312–313.

1985   Tuna and Skipjack Fisheries Yearbook 1985:100–175.

Suisan Shuho Sha

1954   Impacts of the hydrogen bomb experiment over the Bikini Atolls. Fisheries Yearbook 1955:155–167.

Suzuki, Z., et al.

1978   Population structure of Pacific yellowfin tuna. Inter-American Tropical Tuna Commission (IATTC) Bulletin 17(5):277–355.

6.
U.S. Tuna Fleet Ventures in the Pacific Islands

August Felando

BACKGROUND

It was in 1921 that U.S. tuna fishermen first joined forces to pursue their common interests by forming the American Fishermen's Protective Association in San Diego. That organization, after a name change in 1929 and reorganization in 1947, became today's American Tunaboat Association (ATA). In 1952 the ATA merged with the California Tuna Clipper Cooperative to become a nonprofit cooperative association under the terms of California's 1933 Fish Marketing Act.

In 1981 the ATA reorganized again. It suspended its marketing function, which was taken over by the American Tuna Sales Association (ATSA). ATSA represents owners of U.S. purse seiners in marketing frozen tuna.

From its beginning the ATA has been a force in promoting the U.S. tuna industry by fostering the exploration and development of new fisheries, conducting research, improving the technology of fishing, seeking legislation favorable to the tuna industry, and participating in international negotiations for licensing agreements.

Over time the ATA has concentrated its efforts on negotiating sales agreements, collective bargaining agreements with fishermen's unions, and licensing arrangements with foreign governments.

It has testified before Congress for enactment of the Central, Western, and South Pacific Ocean Fisheries Resources Development Act (1947); the Central, Western, and South Pacific Development Act (also 1947); the Tuna Convention Act (1950); the Fish and Wildlife Act (1956); the Fishermen's Protective Act (1967); the Marine Mammal Protection Act (1972); the Atlantic Tuna Conventions Act (1975); and the Fishery Conservation and Management Act (1976). It has been a party to three major initiatives seeking trade relief from foreign tuna imports.

ATA representatives have served as advisers or expert witnesses to U.S. government delegations on bilateral and multilateral agreements, including the Geneva Law of the Sea Conventions (1958 and 1960); negotiations with the governments of Ecuador, Peru, and Chile

(1969–70); the UN Law of the Sea negotiations (1971–82); negotiations with the governments of Mexico and Costa Rica (1977–79); the Eastern Pacific Ocean Tuna Fishing Agreement (1981–83); and negotiations with Pacific island states for a regional tuna treaty (1984–86).

VENTURES IN THE PACIFIC ISLANDS

Pre-World War II (1932–1941)

The U.S. tuna fleet made only two exploratory voyages beyond the eastern tropical Pacific before World War II. In 1932 the San Diego-based tuna clipper *Mayflower*, commanded by Joaquin Medina, completed a fishing voyage of almost 14,400 kilometers. The *Mayflower*, the largest vessel in the fleet at the time, fished around Christmas Ridge and Hawaii. It unloaded only 50 tonnes of yellowfin tuna at Honolulu (Hawaiian Tuna Packers). The crew reported that tuna was abundant but would not take the live bait being used.

A few years later Joe Medina's interest in French Oceania was stimulated by correspondence to the ATA from the French biologist H. Rambke and articles by him in the *Pacific Fisherman* in 1935 and 1936. Rambke claimed that tuna-fishing conditions and oceanographic features of the Marquesas Islands were nearly identical to those of the proven fishing grounds off the Galapagos Islands.

In 1936 Medina, commanding another San Diego-based clipper, the *Cabrillo*, headed for the Marquesas. He left after catching less than 11 tonnes of tuna. The vessel then fished around the Galapagos Islands, where it made excellent catches, before returning to San Diego. This remarkable trip lasted only 57 days.

Although the *Cabrillo's* venture into the Marquesas failed, its success in the waters around the Galapagos Islands showed that these waters were a better gamble for the U.S. tuna fleet. Thus Medina's venture to the Marquesas—3,700 kilometers from San Diego—proved to have important consequences for the development of the central and western Pacific by the U.S. tuna fleet.

World War II (1941–1945)

Following the outbreak of the Pacific war, the U.S. government commandeered 49 San Diego tuna clippers for service in the Pacific. Over 600 fishermen remained with these vessels as crew members and officers. The vessels comprised the greater part of the California tuna fleet, including the largest, most modern ones. Of the 49 tuna clippers, 21, including the *Cabrillo*, were lost in wartime service. Tuna boats, designated "yard patrol" vessels by the Navy, served in most of the major naval engagements in the Pacific, from Midway to Okinawa. Five of the vessels participated in the Guadalcanal (Solomon Islands) campaign. Only one survived. During 1942 and 1943 three former San Diego tuna boats helped transform Funafuti (formerly the Ellice Islands, now the capital of Tuvalu) into a staging base for the Gilbert Islands (Kiribati) campaign and the invasion of Tarawa. The U.S. Navy was so impressed with the design of the tuna clipper that it constructed 30 vessels of the same design. After the war many of these naval vessels were sold to San Diego fishermen who had lost their vessels in wartime service.

In an effort to provide fisheries support for the production of fresh food at island bases in the central and western Pacific, the marine industries section of the U.S. Foreign Economic Administration employed former scientists of Washington's Department of Fisheries—among them Dr. W. M. Chapman, who was appointed ATA's research director in 1951, and Dr. M. B. Schaefer, the noted tuna biologist. During 1943 and 1944 these scientists visited many of the island groups in the central and western Pacific; at many of these places they were able to fish on a semicommercial basis.

In 1946 one of the scientists published a report based on his wartime experiences in the region. He concluded that

> in the vast areas of the tropical Pacific now coming within our sphere of influence the only asset of major importance, aside from national security, is the potential high seas fishery. . . . Tuna of great commercial value appear to be spread uniformly, and in commercial abundance, throughout an area whose size is so great as to be not readily comprehended. . . . The region from the Gilberts east to the Marquesas and south through the Fiji, Cook, Tonga, Samoa, Society and Tuamoto islands have never been exploited commercially for tuna. . . . It is in this great area that the expansion of the American tuna fishery will first enter.

(A few years later the same scientist served as special assistant for fisheries to the U.S. Undersecretary of State. In 1949 he was instrumental in negotiating the convention establishing the Inter-American Tropical Tuna Commission [IATTC], open to signature by governments whose nationals fish for tropical tunas in the eastern Pacific.)

Post-World War (1946–1950)

Tuna fishermen returning to California from the war concentrated on building new tuna vessels or converting naval vessels to civilian use and developing the eastern tropical Pacific tuna fishing grounds as far south as Chile. Between 1946 and 1950 only one exploratory fishing trip was made by the U.S. fleet beyond traditional waters. This trip was made in 1949 by the vessel *Calistar*, which traveled from San Pedro, California, to Christmas Island after obtaining live bait off Mexico, then fished around Fanning, Washington, and Palmyra islands. The vessel caught only 72 tonnes of tuna, and the trip was not a commercial success.

Meanwhile, Hawaiian Tuna Packers (Honolulu) implemented plans to develop new tuna grounds by sending the tuna clipper *Hawaii Tuna* and the sampan *Marlin* to fish south of Hawaii. Hawaiian Tuna Packers had earlier opened negotiations with the French government in Tahiti concerning tuna exploration around the Society Islands. In 1949 the *Hawaii Tuna* made small catches of tuna around the Marquesas, but an inadequate supply of live bait thwarted the venture's success.

Between 1945 and 1948 Hawaiian Tuna Packers also surveyed other central Pacific areas for tuna development. It is believed that the company supported a 1944 proposal for a tuna fishery research program prepared by the Hawaiian Territorial Division of the Department of Fish and Game. Essentially, it was this proposal that became known as the Central, Western, and South Pacific Ocean Fisheries Resources Development Act (1947), or the Farrington Act. Strong support for this research in the central and western Pacific also came from the ATA and other segments of the tuna industry in California, Oregon, and Washington.

Another stimulus to the development of tuna resources in the central Pacific came from a U.S. government agency, the Reconstruction Finance Corporation. In 1947 a private corporation—the Pacific Exploration Company—obtained joint approval from the U.S. Fish and Wildlife Service and the Reconstruction Finance Corporation to operate two vessels—the *Alaska* and the *Oregon*—for a tuna survey in the central Pacific, including the waters around the Line Islands (Kiribati) and the Marquesas. The survey was undertaken in 1948.

The Secretary of the Interior, in remarks at the National Fisheries Institute in New York in 1947, summarized U.S. thinking about exploration and development of the tuna resources in the central Pacific:

> In the investigation of new and undeveloped fisheries, there is a most important field for cooperation between the fishing industry and the federal government. This was brought home to me forcibly during my recent trip to Hawaii and others of our islands in the Pacific. In the areas which I visited, there is evidence of extensive fishery resources—so extensive as to be capable in all likelihood of supplying much more than the existence requirements of the native population. . . . Tuna is a magic word in any community or country which looks to the sea for food and profits. . . . Tunas are unquestionably present in the areas which I visited. The opportunities for development are a challenge to American initiative and the enterprise of the fishing industry.

The secretary assured U.S. industry that his department would provide "at every step the fullest cooperation" in the efforts taken by industry "towards getting the facts it needs for the development of the Pacific fisheries. . . . Every experiment and investigation undertaken by the industry will be facilitated in every way possible by the Department."

Exploratory fishing continued in the central Pacific in 1948. The *Oregon* explored French Frigate Shoals and other islands in the Leeward Island chain and waters around Canton Island (Kiribati) and the Line Islands, particularly in the vicinity of Palymra, Jarvis, and Christmas islands. A research cruise by the vessel *N. B. Scofield* in 1948 also explored tuna fishing areas near Johnston Island.

Other U.S. initiatives in the postwar period that had long-term effects for the U.S. industry were associated with America Samoa and the Trust Territory of the Pacific Islands.

American Samoa. In 1949 an Australian entrepreneur, Harold C. Gatty, established a fishing venture known as South Seas Marine Products. The venture, registered in Fiji, was backed largely by U.S. capital, reportedly from Rockefeller funds. The venture involved deploying three California-style tuna clippers—the *Sea King*, the *Golden Gate*, and the *Rose*—plus two live-bait boats and a refrigerated cargo vessel. The Gatty fleet operated from Fiji, mainly with Fijian crews, primarily because reasonable bait was available there. The company obtained a twin-engined Grumman "Goose" aircraft with a cruising range of 1,900 kilometers to assist in fishing operations. Tuna caught by the fleet was to be transported to American Samoa for processing. An associated company, Island Packers, had obtained permission from the U.S. Navy in 1948 to build a $1.5 million tuna cannery in Samoa. The cannery was constructed, but attempts to supply it with tuna failed, and the plant operated only two trial runs, processing 6 tonnes of fish. The American Samoan government purchased the cannery in 1952 for $40,000. In 1953 the Van Camp Sea Food Company of California bid for, and obtained, a one-year lease to operate it with an option to extend the lease for five years.

(Van Camp still operates a cannery in American Samoa. It now has a processing capacity of 75,000 tonnes of fish a year. Including Star-Kist's production, tuna exports from American Samoa account for more than 90 percent of the territory's total exports.)

Trust Territory of the Pacific Islands. In 1949 the U.S. Navy's deputy high commissioner for the Trust Territory of the Pacific Islands announced that temporary permission for fishing in the territory's waters would be granted, primarily to stimulate exploration of its fisheries.

In response to this initiative, the Japanese undertook nine mothership expeditions in 1950 and 1951. The waters covered by these expeditions ranged between 1 and 6 degrees north latitude and between 137 and 175 degrees east longitude.

Wilvan G. Van Campen, a U.S. government observer on the Japanese vessels, commented on the final Japanese expedition in the Trust Territory and on the future of Japanese tuna fishing in Pacific islands generally. He noted that, as a result of the expeditions, Japanese industry believed the trend in the fishery would be to single-vessel operations involving larger-capacity longliners with greater cruising ranges and improved refrigeration facilities, which could provide high-quality, high-priced fresh fish for an assured Japanese market. The Japanese also believed that the economic and marketing aspects of mothership operations with possible shipboard canning were uncertain and therefore less feasible than longline fishing.

Van Campen, reporting to U.S. government and industry on tuna fishing and canning in American Samoa in 1954, concluded that the Japanese vessels had demonstrated that tuna longlining was viable around the Samoas and that an internationally acceptable pack of canned tuna could be produced in American Samoa.

Van Campen's reports proved prophetic. The success of Japanese longlining, combined with the development of tuna canning in American Samoa, stimulated the U.S. tuna industry and its plans for expansion into the central and western Pacific over the next 20 years.

Pacific Oceanic Fishery Investigation (1949–1959)

The 1947 Farrington Act stated that

> it is the policy of the United States to provide for the exploration, investigation, development, and maintenance of the fishing resources and development of the high seas fishing industry of the United States and its island possessions in the tropical and subtropical Pacific Ocean and intervening areas for the benefit of the residents of the Pacific Island possessions and of the people of the United States.

It was this act that gave rise to the Pacific Ocean Fishery Investigation (POFI) initiative.[1]

Research activities of POFI between 1949 and 1959 concentrated on the waters east of the international dateline and north of 20 degrees south latitude. As a result of knowledge gained during these years and in recognition of the changing conditions of the U.S. tuna industry, POFI reoriented its research programs after 1959. It terminated its exploratory fishing operations in 1960 and directed its research efforts to tuna biology, behavioral studies, and the cultivation of live tuna bait.

Between 1950 and 1951 POFI experimented unsuccessfully with tuna purse seining in the Phoenix, Line, and Hawaiian islands. Longline operations in the same areas proved moderately successful between 1952 and 1954. However, the suggestion that U.S. tuna clippers and seiners convert to longlining found no industry acceptance.

During the early 1950s substantial and adverse changes occurred for the California tuna fleet, largely because of competition from imports of Japanese frozen tuna. By the mid-1950s San Diego's bait fleet had declined sharply in size and suffered financial difficulties.

At the same time, significant technological improvements were being introduced to the San Pedro tuna fleet. Most notably, Larry Zuanich in 1957 converted the first tuna clipper—*Sun King*—to a tuna seiner. Attempts to operate the vessel from San Pedro met with opposition from the fishermen's union, forcing Zuanich to operate from a port in Peru. In 1958 Lou Brito returned to San Diego after an excellent fishing trip with the *Southern Pacific*—the first converted seiner to operate from San Diego.

Thereafter the purse seine revolution took root. Within five years 97 bait boats had been converted to seiners, 3 newly built seiners had entered the fleet, and 8 military vessels had been converted to seiners. It was clear that U.S. industry was turning to purse seining—a fishing method POFI had found unsatisfactory in its experimental programs in the Pacific islands.

Superseiner Years (1969–1980)

The number of U.S. seiners continued to increase through the 1970s, and the trend to larger vessels of the superseiner class escalated. The fleet targeted primarily on traditional tuna fishing areas in the eastern Pacific. Between 1966 and 1970 IATTC annual yellowfin quotas increased by 34 percent, from 79,300 tonnes to 120,000 tonnes (Table 1). However, despite this increase, the length of the agreed fishing season decreased as the closing date was moved back each year, from September 15 in 1966 (258 days) to March 23 in 1970 (82 days). As a result, the U.S. tuna fleet was forced to seek new areas of operation.

In 1967 the fleet had its first success in a new area when three vessels fished off West Africa. This success prompted additional U.S. vessels to move to the South Atlantic in 1968 and 1969. In 1969 two vessels caught tuna in another new area just west of the IATTC's yellowfin regulatory area in the Pacific between Hawaii and California. In 1970 IATTC set the annual yellowfin quota at 120,000 tonnes and the closure date at March 23, further increasing pressure on the U.S. fleet to develop new tuna grounds beyond the eastern Pacific.

Table 1. IATTC yellowfin quotas and fishing season closure dates 1966–1970

	IATTC yellowfin quota in tonnes	Closure date	Days of fishing
1966	79,300	September 15	258
1967	84,500	June 24	175
1968	106,000	June 18	169
1969	120,000	April 16	106
1970	120,000	March 23	82

Source: IATTC. Various years.

Western Pacific. Captain Joe Madruga, who had served during World War II on the *Paramount* in the Solomon Islands and Funafuti campaigns, undertook purse seine surveys in Palauan waters in 1970 at the invitation of the Van Camp Sea Food Company. (Van Camp had established cold storage facilities at Koror in 1964 for the transshipment of tuna landed by Okinawan bait boats owned by Van Camp). Madruga's was the first U.S. tuna venture in the Trust Territory.

After his initial purse seine survey, Madruga persuaded the owners and captains of seven U.S. seiners to participate in exploratory fishing around Palau and then the Philippines. Van Camp and Star-Kist provided some financial support for the exploration. Although fishing around Palau was not successful, the surveys were instrumental in opening the door to the central and western Pacific for the U.S. fleet.

On their way to Palau the seven U.S. seiners—*Mermaid, Kerri M, Connie Jean, Pacific Queen, Conquest, Cabrillo,* and *Polaris*—fished successfully in the eastern Pacific for yellowfin tuna. They diverted to American Samoa to discharge their catches and continued on to Palau. However, the vessels failed to keep their fishing success and the newfound fishing area secret from the San Diego fleet, which began fishing there. Because the fishing was good, four of the exploratory vessels decided to return to the new fishing area in the eastern Pacific. The other three went on to Palau, visited Zamboanga in the Philippines, and also returned to the eastern Pacific fishing area.

Marquesas Islands. In 1971 the U.S. seiner *Kerri M*, under the command of Harold Medina, returned to explore waters around the Marquesas Islands. The vessel took about 230 tonnes of tuna, the first catch of tuna by a U.S. seiner in this area and the largest catch by a U.S. flag tuna boat.

A second exploratory trip to the Marquesas was made by the *Kerri M* in 1972. The trip was partially supported by the U.S. National

Marine Fisheries Service (NMFS), and scientific observers reported on the exploration. Ninety-five tonnes of tuna were taken from a total of 37 sets, 9 of which were successful. After the trip it was apparent that the net being used by U.S. vessels was not well designed for fishing where the water was clear and the thermocline deep. The tuna industry and the NMFS concluded that the net needed to be modified and that it should be patterned after the Japanese nets being used in the central and western Pacific around Papua New Guinea and the Trust Territory. This modification was necessary if U.S. purse seine operations in the area were to be competitive with those of the Japanese fleet.

Pacific Islands Development Commission. In 1969 a paper entitled "Tuna Fishing—A $400 Million Industry for Hawaii, Guam, American Samoa, and the Trust Territory of the Pacific Islands" was prepared by Hawaii's Department of Planning and Economic Development. The paper proposed a series of sea tests using modern purse seine gear and an effort to locate nonseasonal skipjack resources in the central Pacific. In 1970 the ATA and other segments of the U.S. tuna industry were invited to a meeting in Honolulu "to discuss the cooperative effort that should be taken in exploring central Pacific tuna resources." Participants at the meeting included the governors of Guam and American Samoa and the high commissioner of the Trust Territory of the Pacific Islands.

One result of the Honolulu meeting was the creation of the Pacific Islands Development Commission (PIDC), composed of the governors of Hawaii, American Samoa, and Guam and the high commissioner of the Trust Territory. The PIDC was assigned "the priority of the development of the latent skipjack tuna resources of the central and western Pacific." A paper entitled "An American Fisheries Opportunity for U.S. Pacific Islands Peoples" was prepared to reflect the current and potential activities for skipjack development for each of the territories and Hawaii. The paper was distributed to industry, the U.S. Congress, and government agencies.

As a result of the formation of the PIDC, the NMFS started tuna exploration cruises in the western Pacific.

Another outcome of the Honolulu meeting was the 1972 Fong Bill— the Central and Western Fishery Development Act—which authorized the U.S. Secretary of the Interior to carry out a three-year program for "the development of the latent tuna resources of the central and western Pacific Ocean." Between 1970 and 1973 a sum of $3 million was authorized to carry out the purposes of the act. The act was amended several times, but no monies were appropriated by Congress to give it effect.

Pacific Tuna Development Foundation. The Pacific Tuna Development Foundation (PTDF), a private nonprofit Hawaii corporation, was formed in 1974 by government and industry representatives to finance and guide the development of tuna and other fish resources in the Pacific islands region. Its objectives were to assist in the economic development of U.S.-related island areas, to assist the U.S. fishing industry in locating and developing new fishing grounds in the Pacific, and to provide an additional supply of fish for U.S. consumers. PTDF was financed primarily by the Saltonstall-Kennedy Act; strong financial and technical assistance was provided by the U.S. tuna industry.

Between 1974 and 1983 PTDF (later PFDP—the Pacific Fisheries Development Foundation) sponsored 11 exploratory purse seine charters to the central and western Pacific. The first two explorations (1974 and 1976) demonstrated again the need to modify the design of the nets for use in the western Pacific.[2]

New Zealand Operations (1974–1981)

The New Zealand government and Star-Kist Foods (California) entered into an agreement in 1973 to use a U.S. converted purse seiner, the *Paramount*, to survey the country's pelagic fish resources. In 1974 and 1975 the vessel, under the command of Anthony Tipich of San Pedro, explored New Zealand waters and caught skipjack tuna. This exploration became the springboard for the development of the fishery in the western Pacific south of Guam for the U.S. tuna purse seine fleet. In six seasons the U.S. fleet took about 44,000 tonnes of skipjack in New Zealand waters. Indeed, it was the New Zealand fishery that made it possible for U.S. seiners to gamble on further ventures west of the international dateline.

Western Pacific (1979–1981)

Factors stimulating the expansion of the U.S. fleet into the western Pacific were the 1974 purse seine explorations in Papua New Guinea waters, subsequent PFDF exploratory charters, and the development of the skipjack fishery in New Zealand.[3]

The fleet showed its desire for long-term commitments to operations west of the dateline in 1980, when representatives of the ATA entered into negotiations with the governments of Palau, the Federated States of Micronesia, and the Marshall Islands to conclude a fishing agreement.[4]

These negotiations followed an approach by Palau to the ATA in 1979.[5] Palau had sought to discuss a fisheries agreement patterned after agreements worked out with Japan's tuna fisheries associations. The ATA responded by forwarding a draft proposal for a multilateral arrangement between the ATA and the maritime authorities of Palau, Micronesia, and the Marshall Islands. An agreement was concluded for the period July 1, 1980, to June 30, 1982.

This regional licensing arrangement, the first of its kind, represented a decision by some elements of the U.S. tuna industry to stake their future in the western Pacific. Lou Brito, manager of Ocean Fisheries, expressed his confidence by having the governor of American Samoa name his new Pago Pago-based vessel the *Tifaimoana*. Larry Zuanich and his two sons decided to make Guam the home port for seven tuna seiners operated by Zee Enterprises and to develop Tinian (Commonwealth of the Northern Marianas) as a transshipment base for tuna. Brito and Zuanich had been the first in the U.S. fleet to convert bait boats to purse seiners in 1957 and 1958. Twenty-two years later they again led the industry in spearheading a major change by moving vessels from the traditional eastern Pacific grounds to the western Pacific grounds.

CONCLUSION

This paper presents an account of the U.S. tuna industry's involvement in the central and western Pacific and its part in developing the region's tuna fishery from 1932 to the conclusion of ATA's first agreement in the Pacific islands region in 1980.

No attempt has been made here to record the contributions of U.S. tuna canners to industry development in the region over the last 60 years. For example, in 1928 the San Diego-based tuna bait boat, *Del Monte*, under the command of Manuel Freitas, headed for the Philippines with the goal of catching tuna for a company financed by the California Packing Corporation of San Francisco. This venture is believed to be the first tuna investment made by a U.S. corporation in the western Pacific.

It is hoped that this paper will stimulate more careful and exhaustive research to document the role of the U.S. tuna industry in developing the central and western Pacific tuna fishery.

NOTES

1. In 1948, explaining the work of POFI, Milner B. Schaefer, biologist for the U.S. Fish and Wildlife Service, told the U.S. fishing industry:

    > Studies of the Pacific Oceanic Fisheries Laboratory will be concentrated on the tuna. These valuable oceanic fishes offer the greatest possibilities for the development of valuable commercial fisheries. . . . Greatly expanded studies of all the factors in the life history and behavior of the tunas will be possible.

It is important to find out about their migrations, feeding habits and depth distribution at an early date in order to assist in the development of profitable fisheries. Of longer range importance is the problem of whether or not the tuna of Oceania and the seas off southern California, Mexico, central America and south America are all one stock. If so, the development of new fisheries in Oceania will be in direct competition with the existing west coast fishery. If not, the development of the new fishery will supplement rather than compete with the west coast fishery.

In 1951 Schaefer, as IATTC's director of investigations, concluded on the basis of scientific studies that the yellowfin tunas of Oceania and the eastern tropical Pacific were of different stocks.

2.   Other tuna-related projects sponsored by PTDF/PFDF between 1974 and 1985 were (1) tuna trolling in the Federated States of Micronesia (1977-78); (2) live-bait tuna operations and bait surveys in the Federated States of Micronesia (1980); (3) exploratory longline fishing south, west, and north of Hawaii (1980); (4) design, construction, deployment, and monitoring of anchored fish aggregating devices (FADS) in Hawaii, American Samoa, Palau, Guam, and the northern Marianas (1977-81); (5) culture and test-fishing of bait fish in Hawaii (1977-85) and in American Samoa (1973-78); (6) transportation of live anchovies from California to Hawaii (1976-77); (7) aerial spotting of skipjack off Hawaii (1977); (8) feasibility study of an *ika shibi* fishery for Guam (1979) and night tuna handling in the Federated States of Micronesia (1978); (9) trolling in the Marshall Islands (1980-81); (10) survey of Palau's bait-fish resource (1980-81).

3.   U.S. commercial fishing operations started in Papua New Guinea, Palau, and the Federated States of Micronesia in 1979.

4.   ATA's subsequent agreements in the region were with Papua New Guinea (1982); Federated States of Micronesia, Kiribati, and Palau (1983-84); and the Cook Islands, Niue, Tokelau, Tuvalu, and Western Samoa (1983-84).

5.   This approach was made by Alan B. Chapman, director of Palau's Maritime Authority. He was the son of Dr. W. M. Chapman, who had been closely associated with the ATA in the 1950s. Alan Chapman subsequently worked for the Forum Fisheries Agency.

REFERENCE

Inter-American Tropical Tuna Commission.
   annual   Various years. La Jolla, California.

7.
U.S. Tuna Policy: A Reluctant Acceptance of the International Norm

Jon M. Van Dyke and Carolyn Nicol[1]

INTRODUCTION

Some of the world's most productive tuna fishing grounds are within 200 miles of the coasts of newly independent Pacific island nations. U.S. purse seine vessels have persistently disregarded these nations' claims to jurisdiction over the tuna in the region. The resulting "tuna wars" in which boats were seized for poaching—combined with a recognition that important security issues were also at stake—created the environment in the mid-1980s within which the United States negotiated a regional fishery agreement. Signed in early 1987, the agreement implicitly recognizes the rights of the island nations to the tuna resource. It also establishes a program of substantial payments that should help economic development in the region. With this change, U.S. policy on tuna has been brought into line with the policy of the rest of the international community.

U.S. POLICY AND THE LAW OF THE SEA

Two philosophies about ownership of tuna and tunalike fish came into conflict during the 1974–1982 negotiations of the Third United Nations Conference on the Law of the Sea (UNCLOS III). The United States argued that highly migratory species are not the property of any nation because they move from one region to another. To manage the resource effectively, the United States argued, coastal nations should work with distant-water fishing nations to develop a regional approach to conservation and management. But developing coastal nations argued that they had sovereign rights over the migratory species while the fish were within their 200-mile exclusive economic zones (EEZs) and that no other nation had a right to participate in decisions affecting the resource while it was in the EEZ (Joseph 1977, 280). Other distant-water fishing nations agreed with the U.S. position at first. By the late 1970s, however, the United States was virtually the only nation still arguing that coastal nations did not have exclusive rights over highly migratory species while they were within their EEZs.

The U.S. "species" approach to jurisdiction over fish was consistent with earlier U.S. legislation encouraging U.S. distant-water fishing vessels to "violate" the claims of coastal nations and to protest their assumption of jurisdiction over tuna resources (Bugosh 1984, 2–3). Seizure of U.S. tuna boats did not deter the "poaching" because the U.S. government reimbursed all expenses incurred by "illegally" seized U.S. ships under the provisions of the 1967 Fishermen's Protective Act (U.S. Code, vol. 22, sec. 1971 et seq.). Whatever penalty a coastal nation imposed on a shipowner was in fact paid by the U.S. government, which then subtracted the sum from any foreign aid that would otherwise have gone to that nation (U.S. Code, vol. 22, sec. 1975).

The negotiating texts developed at UNCLOS III all called for international and regional cooperation to guarantee effective conservation and optimum use of highly migratory species, but these same texts also granted coastal nations "sovereign" rights to exploit and manage all living resources within their EEZs (Van Dyke & Heftel 1981, 58). Article 56 in the 1982 Law of the Sea Convention established that coastal states have "sovereign" rights over all resources within their exclusive economic or fishery zones. Article 64, on highly migratory species, does, however, require coastal nations to "cooperate directly or through appropriate organizations with a view to ensuring conservation and promoting the objective of optimum utilization of such species throughout the region, both within and beyond the exclusive economic zone." (See Appendix 1). This language permitted the United States to maintain its legal position, even though it stood alone in denying that the "sovereign" rights of the coastal nations included the highly migratory species found in their EEZs. By the mid-1970s, pressure from U.S. coastal fishing interests seeking protection of their fishery resources led to the passage by Congress of the 1976 Fishery Conservation and Management Act (FCMA) and the acceptance by the U.S. government of the concept of the 200-mile EEZ. Accommodating the interests of two segments of the U.S. fishing industry—domestic fishing in U.S. coastal waters and long-distance fishing by U.S. tuna boats in foreign nations' coastal waters—created legislative inconsistencies. The FCMA established a 200-mile fisheries zone, but it also created an exception for "highly migratory species" and took a unique approach to defining this term.

THE 1976 MAGNUSON FISHERY CONSERVATION AND MANAGEMENT ACT (FCMA)

The Fishery Conservation Zone

The FCMA established a "fishery conservation zone" within 200 miles of all U.S. coastlines—the Atlantic, the Pacific, the Gulf of Mexico, Alaska, and all the U.S. entities in the Pacific. Congress presented as the rationale for this bill that foreign overfishing in U.S. coastal waters was a serious problem: "Many coastal areas are dependent upon fishing . . . and their economies have been badly damaged by the overfishing of fishery resources. . . . The activities of massive foreign fishing fleets . . . have contributed to such damage [and] interfered with domestic fishing efforts" (U.S. Code, vol. 16, sec. 1801(a)(3)); and "a national program for the conservation and management of the fishery resources of the United States is necessary to prevent overfishing, to rebuild overfished stocks, to insure conservation and to realize the full potential of the Nation's fishery resources" (U.S. Code, vol. 16, sec. 1801(a)(6)).

Highly Migratory Species and the FCMA

To protect the U.S. distant-water fishing fleet's right to claim that no coastal state—including the United States—has jurisdiction over highly migratory species, the FCMA created a special exception for tuna. Congress recognized that "highly migratory species of the high seas . . . constitute valuable . . . natural resources [which] contribute to the food supply, economy and health of the Nation" (U.S. Code, vol. 16, sec. 1801(a)(1)). Within the U.S. fishery conservation zone, however, any nation can fish for tuna because the United States declines to exercise exclusive authority over tuna. Under the special exception, "the exclusive fishery management authority of the United States shall not include, nor shall it be construed to extend to, highly migratory species of fish" (U.S. Code, vol. 16, sec. 1813).

The term *highly migratory species* is defined as "species of tuna which, in the course of their life cycle, spawn and migrate over great distances in water of the ocean" (U.S. Code, vol. 16, sec. 1802(14)). It is widely held that the U.S. definition equating highly migratory species with tuna is inconsistent with scientific definitions, and it is certainly inconsistent with Annex 1 of the 1982 United Nations Law of the Sea Convention, which lists 17 categories of highly migratory species, including marlin, sailfish, and swordfish in addition to tuna (Harrison 1985, 352). The FCMA definition also overlooks some species, such as wahoo, bonito, and pilchard, which are migratory but are not listed in the Convention (Burke 1985, 330).

The tuna exception was included to benefit the U.S. distant-water fishing fleet, but it has had an adverse effect on some regional fishing interests within the United States. In 1981, 11 U.S. senators introduced the American Tuna Protection Act in an unsuccessful attempt to amend the FCMA by deleting the exception for highly migratory species. One senator, describing the damaging effect of Japanese longline fishing efforts in the Atlantic on New England's domestic bluefin tuna industry, characterized the tuna exception as a "serious gap in the overall fishery management program of this Nation" (Weicker 1981, S9107). U.S. bluefin fishermen in the Gulf of Mexico and U.S. tuna fishermen in the Northern Marianas, Guam, American Samoa, and Hawaii also suffer because of the FCMA, and local fishing interests in these regions could benefit from U.S. control over tuna in the U.S. EEZ (Harrison 1985, 353–54).

The Deputy Assistant Secretary of State for Oceans and Fisheries Affairs acknowledged in 1981, in testimony on the proposal to delete the highly migratory species exception from the FCMA, that the U.S. refusal to recognize jurisdiction over tuna is designed to support its distant-water fishing fleet in challenging other nations' claims:

> The tuna exclusion in the FCMA is both symbolic and practical evidence of U.S. Government support of our tuna industry. . . . The embargo provisions of the [FCMA] and the compensation programs of the Fishermen's Protective Act . . . provide the U.S. with important negotiating leverage and without them, the U.S. tuna fleet would be at the mercy of . . . coastal States in the 200-mile zones. . . . Without these legal protections, the U.S. tuna fleet would soon decide that there is no future in attempting to operate on a commercially viable basis under the U.S. flag. U.S. tuna vessels would be compelled to buy very expensive coastal State licenses and to adhere to whatever arbitrary or discriminatory regulations were imposed by the coastal States, or indeed to remain outside certain zones entirely, without regard to the location of this rich resource. (Kronmiller 1981, 8)

International Fishery Agreements

The provision that has caused the most difficulty in U.S. relations with the Pacific island nations is U.S. Code, vol. 16, sec. 1822, International Fishery Agreements. It requires coastal nations to join with distant-water fishing nations (meaning in particular the United States) to manage the highly migratory species through an international fishery agreement in order to gain U.S. recognition of their fishery conservation zones:

> NONRECOGNITION—It is the sense of the Congress that the United States Government shall not recognize the claims of any

foreign nation to a fishery conservation zone . . . beyond such nation's territorial sea . . . if such nation . . . fails to recognize and
accept that highly migratory species are to be managed by applicable international fishery agreements. . . . (U.S. Code, vol. 16, sec.
1822(e))

The United States will not recognize the tuna management authority
of an international fishery agreement unless the United States is a party
to the agreement. Nor will the United States recognize a fishery conservation zone if the nation claiming the zone fails to "take into account
traditional fishing activity" of U.S. fishing vessels (U.S. Code, vol. 16,
sec. 1822(e)(1)) or if that nation imposes conditions "unrelated to fishery conservation and management" on American fishing vessels (U.S.
Code, vol. 16, sec. 1822(e)(3)).

Sanctions: Import Prohibitions

If a U.S. tuna boat is seized in another nation's fishery zone beyond
the territorial sea "as a consequence of a claim of jurisdiction which
is not recognized by the United States" (U.S. Code, vol. 16, sec. 1825
(a)(4)(c)), the Secretary of State notifies the Treasury Department to take
"necessary and appropriate action" to embargo that nation's fish
products. Similarly, if a coastal nation fails to negotiate in good faith
with the Secretary of State to bring about "an international fishery agreement allowing fishing vessels of the United States equitable access to
fisheries . . . in accordance with traditional fishing activities of such
vessels" (U.S. Code, vol. 16, sec. 1825(a)(1)), the Secretary of State again
notifies the Secretary of the Treasury (U.S. Code, vol. 16, sec. 1825(a)).
Section 1825(b) (Prohibitions) details the response expected from the
Secretary of the Treasury:

> Upon receipt of any certification from the Secretary of State . . .
> the Secretary of the Treasury shall immediately take such action
> as may be necessary and appropriate to prohibit the importation
> into the United States—(1) of all fish and fish products from the
> fishery involved, if any; and (2) upon recommendation of the Secre-
> tary of State, such other fish or fish products, from any fishery of
> the foreign nation concerned, which the Secretary of State finds
> to be appropriate to carry out the purposes of this section. (U.S.
> Code, vol. 16, sec. 1825(b))

Albacore tuna, for example, might be the "fishery involved" if a
U.S. tuna boat were seized in the North Pacific. On the recommenda-
tion of the Secretary of State, other types of fish produced by the seiz-
ing nation may also be prohibited from importation. *Fish* is defined in
section 1825 as "any highly migratory species" (U.S. Code, vol. 16, sec.
1825(d)(1)), and *fish products* means "any article which is produced from

or composed of (in whole or in part) any fish" (U.S. Code, vol. 16, sec. 1825(d)(2)).

Are the Sanctions Mandatory?

Although the application of sanctions under the FCMA has been generally believed to be mandatory, a 1986 U.S. Supreme Court decision suggests that it may not be. In this case, the Court considered whether marine mammal provisions of the FCMA required the U.S. government to impose sanctions against Japanese whaling vessels violating harvesting limits (Japan Whaling Association v. American Cetacean Society 1986). The Court agreed with the U.S. Secretary of Commerce that the statutory language permitted him to exercise discretion and to refrain from certifying that Japan had "diminish[ed] the effectiveness of an international fishery conservation program."

The whaling certification procedure is similar to the Secretary of State's certification to the Secretary of the Treasury, under U.S. Code, vol. 16, sec. 1825 (described above), that a nation is not part of an international fishery agreement on highly migratory species. For marine mammals, the Secretary of Commerce communicates directly to the President, who may choose to direct the Secretary of the Treasury to impose sanctions. For highly migratory species, the Secretary of State notifies the Secretary of the Treasury directly that a U.S. vessel has been seized or that U.S. vessels are being improperly excluded from a traditional fishing ground.

It could be argued that the ambiguous phrase "take such action as may be necessary and appropriate to prohibit importation" (U.S. Code, vol. 16, 1825(b)(1)) empowers the Treasury Department to withhold sanctions. The contrary argument is that the wording clearly requires the Secretary of the Treasury to take some action ("shall immediately take . . . action") and that the phrase "such action as may be necessary and appropriate" creates discretion as to means but does not diminish the obligation to act "to prohibit importation" of fish products.

Although the consistent practice of the executive branch in the vessel-seizure incidents described below lends support to the argument that its duty to impose sanctions is automatic and mandatory rather than discretionary, the American Cetacean decision and the language of the FCMA permit the interpretation that the executive branch may exercise discretion in determining whether to impose sanctions. The statute provides no clear guidelines for the State Department to follow in certifying to the Secretary of the Treasury that, for example, a nation has "refused to commence" negotiating an international agreement or has not negotiated "in good faith." And, although nothing in the FCMA language suggests that the State Department has discretion in determining that a U.S. vessel has been seized, the State Department, in what appears to be an exercise of discretion, has refrained from taking action against a nation until the boat owner requests compensation (Post-Courier 1982, 2). Thus the Cetacean decision supports a conclusion that the role of the executive branch is discretionary and that the FCMA sanctions need not be applied automatically as each incident occurs.

U.S. TUNA EMBARGO ON CANADA

In late August 1979, 19 U.S. tuna vessels fishing for albacore tuna within Canada's 200-mile fisheries zone were seized by Canada. The United States swiftly imposed an embargo on all Canadian tuna and tuna products, effective August 31, pursuant to U.S. Code, vol. 16, sec. 1825 of the FCMA (Doulman 1986b, 1). The chairman of the Senate Foreign Relations Committee later explained:

> As a consequence of the Canadian seizures of U.S. flag vessels, the United States imposed an embargo on imports of tuna and tuna products from Canada as required by the Fishery Conservation and Management Act. Canada has challenged this embargo as a violation of our international obligations under the General Agreement on Tariffs and Trade (GATT) and the issue is currently under consideration by a GATT Panel in Geneva. (Percy 1981, 16462)

Of the total U.S. imports of tuna and tuna products, less than 1 percent came from Canada (GATT panel 1981, 8–9). The embargo was lifted as a result of an interim agreement, and then the Treaty on Pacific Coast Albacore Tuna Vessels and Port Privileges was negotiated to resolve the underlying dispute over fishing areas (Gorton 1981, 16462). Although the embargo was lifted, Canada asked the GATT panel to evaluate Canada's claim that the United States had violated the GATT. The panel agreed with Canada that "the United States prohibition of imports of tuna and tuna products was clearly inconsistent with the obligations of the United States under Article XI(1) not to institute or maintain prohibitions other than duties, taxes or other charges on the importation of any product of the territory of any other contracting party." The U.S. embargo was found inconsistent with Article XX(g) because the United States did not sufficiently show the embargo to be related to domestic production or consumption, a requirement of Article XX(g) (GATT Panel 1981, 7, 26).

SEIZURES IN THE PACIFIC

Since 1982, four U.S. tuna boats have been seized in the Pacific islands: the *Danica* in Papua New Guinea in 1982, the *Jeannette Diana* in waters off the Solomon Islands in 1984, and the *Ocean Pearl* and the *Priscilla M* in the Federated States of Micronesia in 1985 and 1986.

Papua New Guinea

On February 10, 1982, a Guam-based U.S. purse seiner, the *Danica*, carrying 600 tons of tuna, was arrested by the Papua New Guinea defense force 35 miles inside its 200-mile economic zone (McCloskey 1982, 3947; Pacific Islands Monthly 1982, 51). A senior Papua New Guinea fisheries inspector described the incident:

> At 0930 hours, Wednesday, February 10th, I departed Rabaul on board a P.N.G. Defense Force aeroplane to carry out a patrol in the Lyra Reef area of the Declared Fishery Zone. . . . At 1134 hours an echo was picked up on the aeroplane's radar and upon closing I could see the echo was a purse seiner type foreign fishing vessel. I plainly read the vessel's name DANICA painted on the bow. On the stern were the words DANICA and SAN JUAN P.R. . . . As the foreign fishing vessel was carrying out fishing operations, was inside P.N.G.'s 200-mile Declared Fishing Zone and had no license to fish in P.N.G.'s Declared Fishing Zone, I informed Major Bawden that she should be arrested and brought to Rabaul for further investigation. (Coase 1982, 2)

When the *Danica* reached Rabaul on February 12, its captain refused to answer questions. However, a Papua New Guinea Defense Force officer boarded the *Danica* on February 11 and later described the event:

> I asked the Captain how he came up to be here. I was told that as they normally fish off logs and due to the drift of the logs they were fishing on, [the vessel] drifted into our water. I asked the captain [about] Navigational aid on the ship and he told me that he has Satellite Navigation equipment [so] he himself knew he was fishing in our water. (Kager 1982, 2)

An editorial in a Papua New Guinea newspaper called the United States "a firm and long time friend of Papua New Guinea" (Times 1982, 32), but residents subsequently expressed resentment toward U.S. fishing in letters to the editor: "My admiration for the greatest nation in the world today—the United States—is gradually waning. [I]f I were in Yankee land . . . I wouldn't even dream of trespassing on dear ole Billy Carter's farm" (Lebasi 1982, 2). Another writer portrayed the United States as the "Big Boy," a "kid on the block, much bigger than the other kids, and he takes things from them . . . and nobody says much because he's big and mean, and it's not real smart to cross a guy like that" (Persevero 1982, 2).

The prime minister of Papua New Guinea resolved not to "bow down to threats," such as an embargo on tuna products, saying his "primary concern in this matter is to make all countries respect P.N.G.'s laws, and that includes the United States" (Laumaea 1982a, 1). The prospect of a retaliatory embargo under the FCMA created an uncomfortable situation: a fishing embargo would cause Papua New Guinea an annual loss of about $20 million, and Washington did not want to penalize a nation largely dependent on foreign aid (Trumbull 1982, B-3).

Before the embargo took effect, the *Danica* was resold to its owner for $270,000 (PNGK200,000), an estimated 5 percent of its assessed value (Laumaea 1982b, 1). U.S. tuna boat owners agreed to purchase fishing licenses from Papua New Guinea for $35 per ton of vessel weight (Trumbull 1982, B-3). The per-ton rate schedule was patterned after a two-year agreement between the American Tunaboat Association (ATA) and the U.S. Trust Territories in the Pacific (Marshall Islands, Palau, and the Federated States of Micronesia); the dollar value was related to the success of competitors from Japan and Korea also fishing in Papua New Guinea's waters, according to the American Tunaboat Association (Niugini News 1982, 5).

Both in Papua New Guinea and in the United States, the *Danica* left ruffled feathers. In Papua New Guinea, the Prime Minister's agreement was characterized as a "sellout" because (1) the purchase price was a small fraction of the vessel's worth, and (2) an agreement reached only with the American Tunaboat Association—and not the U.S. government itself—fell short of the target: official U.S. government recognition of Papua New Guinea's 200-mile declared fishery zone (Diro 1982, 12).

In the U.S. House of Representatives, inconsistencies in U.S. policy were pointed out: "It is time to abandon our contention that we can control fishing in our own 200-mile zone while still claiming the right to fish for tuna within other nations' 200-mile zones" (McCloskey 1982, 3947). The representative further noted that the United States requires foreign fishing vessels to obtain licenses to fish for all resources other than tuna in U.S. fishery conservation zones and added:

> Papua New Guinea is properly incensed over the combination of our tuna industry's refusal to abide by Papuan New Guinea law and the resulting imposition of U.S. sanctions. . . . [It] seems arrogant in the extreme for us to deny small countries the right to [protect an area similar to our own 200-mile conservation zone]. . . .
> A continuance of strong-arm tactics against a small country such as Papua New Guinea may endanger long-standing friendships far

more valuable to us than a slight increase in tuna costs through acceptance of licensing fees by our tuna industry. (McCloskey 1982:3948)

Solomon Islands

In June 1984, another U.S.-registered seiner, the *Jeannette Diana*, was arrested for illegal fishing in the 200-mile zone of the Solomon Islands. The High Court of the Solomon Islands fined the captain and owner a total of $50,400 (SI$72,000) (Keith-Reid 1984, 10). More significantly, the court also ordered the boat and its catch, fishing gear, and helicopter to be forfeited to the Solomon Islands and put up for sale. The owners paid $539,000 (SI$770,000) to buy it back, but under the Fishermen's Protective Act, the $539,000 was recouped from potential U.S. aid to the Solomon Islands.

An FCMA embargo on all tuna and tuna products was imposed on August 23, 1985 (Doulman 1986b, 1). On August 27, the Ministry of Foreign Affairs in Honiara responded in a news release saying that the Solomon Islands government would reconsider a Soviet application to fish in the country's EEZ. The news release also stated that no fishing licenses would be issued to U.S.-owned or -operated boats, and U.S. fishing vessels would be banned from entering Solomon Islands waters during the embargo (Tsamenyi 1985, 38).

Federated States of Micronesia

On July 24, 1985, the U.S. fishing vessel *Ocean Pearl* was discovered fishing in the EEZ of the Federated States of Micronesia. When authorities approached, the vessel's captain expressed an intention to drop a speedboat rigged to a boom onto the boarding party (National Union 1985, 5). The Federated States of Micronesia applied to remove the *Ocean Pearl* from the Forum Fisheries Agency's (FFA's) Regional Register of Fishing Vessels—a removal which would have meant that this vessel could not fish in the 200-mile zone of any Pacific island nation. The boat owners brought the vessel to Honolulu, where it was repainted and renamed. The Federated States of Micronesia government persisted in seeking to have the vessel removed from the Regional Register and in pressing court charges. This persistence induced the boat's owners to pay the Federated States of Micronesia $500,000 to refrain from removing the vessel from the Regional Register, thus settling the dispute (Doulman 1986d).

On August 14, 1986, a Guam-based U.S. seiner, the *Priscilla M* (the sister ship of the *Jeannette Diana*), pulled into Pohnpei to seek medical attention for an injured crew member. Government officials inspecting the vessel's logbook discovered evidence that the *Priscilla M* had been fishing extensively in the 200-mile zone of the Federated States of Micronesia and had even fished in the 12-mile territorial sea adjacent to the state of Yap on several occasions since November 1985. The boat was promptly seized, and civil and criminal charges were filed. The captain and crew were delayed a full month until an out-of-court settlement could be negotiated between the boat owners and government officials. Pursuant to this settlement, the owners paid a $400,000 fine for fishing violations and agreed to pay $58,000 for a license fee. The United States also agreed not to impose economic sanctions in response to the seizure (King 1986, 3). (It is likely that both sides moderated their demands in this incident in order not to disrupt the ongoing regional treaty negotiations that appeared to be nearing a conclusion.)

JUSTIFYING U.S. FISHING IN OTHER NATIONS' EXLUSIVE ECONOMIC ZONES

International Cooperation and Article 64

In 1983, the Assistant Secretary of State for Oceans and Fisheries Affairs described the U.S. State Department position as follows:

> The rationale behind the United States approach is straightforward. Tuna are not a resident resource of the EEZ. They are only found within any EEZ temporarily and may migrate far out into the ocean waters beyond. Therefore, the coastal state does not have the ability to manage and conserve tuna, nor does it have a paramount interest in their development. Although many coastal states claim jurisdiction over tuna within 200 nautical miles, none exercise conservation and management authority through purely domestic measures. Only through international agreements have States actually managed effectively the highly migratory tuna species. . . . Accordingly customary international law precludes the coastal state from establishing sovereign rights over tuna. In the U.S. view this is evidenced by Article 64 of the Law of the Sea Convention, which requires cooperation between coastal states and distant-water fishing nations to manage tuna, both within and outside the EEZ, on a regional basis, through an international organization. It is the view of the United States that Article 64 precludes the coastal state from establishing sovereign rights over tuna. (Kronmiller 1983, 2)

The United States has supported this argument by pointing out that a coastal state's authority over resources in its EEZ is subject to "other rights and duties provided for in this convention" (Article 56(c)). The language of Article 64 ("shall cooperate directly or through appropriate international organizations") arguably imposes on coastal nations a mandatory duty to cooperate with foreign fishing nations to ensure "conservation" and to promote "optimum utilization" of the species.

The official U.S. position is that Article 64 thus precludes exclusive coastal state jurisdiction over tuna (Burke 1984, 308). No other nation supports the United States on this issue (Burke 1984, 304).

International Fishery Agreements and the FCMA

The United States also argues that it is bound by the FCMA to have a regional agreement. The FCMA prohibits the United States from recognizing a nation's fishing zones "if that nation fails to recognize and accept that highly migratory species are to be managed by applicable international fishery agreements" (U.S. Code, vol. 16, sec. 1822). The FCMA defines an *international fishery agreement* as a fishing agreement, convention, or treaty to which the United States is a party (U.S. Code, vol. 16, sec. 1803 (15)). Thus, to satisfy the FCMA, the United States must be a party to an agreement, and the highly migratory species must be managed by the organization established by the agreement. The legal position advanced in this U.S. legislation requiring international management of tuna is not supported by other nations; it has been opposed in particular by the FFA, the organization established by the independent Pacific island nations to coordinate tuna policies (Van Dyke & Heftel 1981, 47, 58).

Role of the U.S. Tuna Industry

The ATA and the United States Tuna Foundation (USTF) are the two major industry lobbying groups that pressure Washington policy makers. These groups portray tuna fishing as a marginal industry that has taken a beating in recent years and must therefore keep its costs down. Because President Reagan's economic policy favors self-reliance and discourages government subsidies to floundering enterprises, his administration has faced an ideological dilemma on this issue. The true laissez-faire approach would be to let the industry sink or swim on its own, but the Reagan administration has supported the industry's efforts to obtain financial support. Industry lobbyists have been so persuasive that the administration has provided what appears to be a large subsidy in the regional fishing treaty concluded in early 1987.

The U.S. tuna fleet has moved steadily during the 1980s from the eastern to the western Pacific. In 1979, the president of the ATA said that only three U.S. seiners were operating in the western Pacific, that the most there had ever been was five, and that the United States posed no threat to developing nations' efforts to control fishery resources in their own EEZs (Van Dyke & Heftel 1981, 15). But by 1982, the situation had changed dramatically, and the president of the ATA told the House Committee on Merchant Marine and Fisheries that "many U.S. vessels are moving their operations to the Western, Central, and

Southern Pacific, where a tremendous potential in skipjack is available. As many as 23 vessels, or 20 percent of the total U.S. tuna fleet, will operate exclusively in this area this year" (Felando 1982, 4). By 1984, 69 U.S. seiners fished in the western Pacific, and 66 percent of the total U.S. catch came from the region (Herrick and Koplin 1984, 3). In May 1986, the ATA reported that the U.S. tuna seiner fleet was "about equally divided operationally between fishing grounds in the Eastern Pacific and the Western Pacific" (Felando 1986, 2). The two reasons for this shift of fishing effort to the western Pacific were the difficulty in concluding eastern Pacific fishing negotiations concerning rights to fish in waters off Mexico and Costa Rica (Felando 1982, 10), and El Niño, the periodic shift of current that causes water temperatures and fishery patterns to change (Islands Business 1986a, 26).

NEGOTIATIONS

The U.S. legislative process and domestic political pressures make it difficult to dismantle a law once it is in place. Instead of working to repeal the FCMA, officials attempted to negotiate an international fishery agreement designed to meet the letter (if not also the spirit) of the FCMA. A U.S. diplomat explained this strategy while the negotiations were in progress:

> The very fact that we are conducting these negotiations indicates we recognize that [changing the law] is a problem. In other words, what we are offering to do is to satisfy the requirements of the countries for control over the tuna fish assets in their EEZs without violating our own law. (Islands Business 1986b, 14, quoting Edward Dillery, U.S. Ambassador in Fiji)

In early 1987, after ten rounds of negotiations over a two-year period, the United States finally came to an agreement wih the 16 Pacific island nations participating through the FFA on a multilateral fisheries treaty (Honolulu Advertiser 1986a, D-1). The principal issues were (1) licensing and the Regional Register of Fishing Vessels, (2) closed and limited areas, (3) catch reporting, and (4) licensing fees. Many other less politically charged "technical" issues (such as how to identify fishing vessels) were also discussed.

The Deputy Assistant Secretary for Oceans and Fisheries Affairs was the chief negotiator for the United States. The ATA president usually attended in an advisory capacity. Beginning in round five, a senior State Department counselor, Edward Derwinski, also participated in the negotiations because strategic considerations arose, triggered by Kiribati's agreeement in August 1985 to allow up to 16 Soviet fishing vessels to fish in its EEZ for 12 months for a fee of $1.5 million (Doulman

1986a, 5). One Pacific commentator reported in November 1985 that the Kiribati-Soviet agreement

> was a development that the Americans, anxious to keep the Pacific islands clear of any Russian influence, had been dreading. It spurred them to work harder for a multilateral treaty with the Forum Fisheries countries. At previous talks the Americans had displayed what island delegations initially described as a good deal of ignorance, insensitivity and more than a touch of arrogance concerning Forum Fisheries aspirations. But at the first four rounds of talks some progress towards a treaty was made. The U.S. delegation was led by a new face, Ed Derwinski, sent by Secretary of State George Shultz specifically to confound the Russians by getting a treaty concluded as fast as possible. Island delegates told Islands Business that they had never found the American side more amenable. (Islands Business 1985, 27)

Pacific Island Representatives

The 16 Pacific island participants were Australia, the Cook Islands, Federated States of Micronesia, Fiji, Kiribati, Marshall Islands, Nauru, New Zealand, Niue, Palau, Papua New Guinea, Solomon Islands, Tonga, Tuvalu, Vanuatu, and Western Samoa. Before each negotiating session with the United States, island representatives met to develop a common negotiating position. The spokesperson for the islands' position was chosen on a rotating basis; the United States did not know in advance which nation would represent the islands. The staff of the FFA provided technical and research support, and the director of the FFA usually chaired prenegotiation meetings (Doulman 1986a, 2).

Licensing and the Regional Register

In the fourth round the United States, fearing arbitrary denial of a license, asked that the treaty spell out criteria for issuing fishing licenses. The islanders, wanting to maintain discretion over issuing licenses, insisted that all vessels be in "good standing" on the Regional Register of Fishing Vessels prior to issuance of a license. A vessel would not be in "good standing" if it had violated any rule of any nation participating in the FFA. U.S. negotiators felt that arbitrary application of rules might lead to blacklisting. The FFA administers the Regional Register, which has proven to be a potent weapon against illegal fishing. The consequences of violating the rules are so grave that, even though the chances of getting caught are slight, the Regional Register tends to be a self-enforcing mechanism. The United States also wanted the regional agreement to set forth all terms and conditions relevant to U.S. fishing to avoid subjecting U.S. tuna boats to a multiplicity of coastal state laws

(Wolfe 1985, 3). In the sixth round, at Kona on the Big Island of Hawaii, the United States agreed that U.S. vessels would be subject to the Regional Register (Wolfe 1986a, 1).

Closed and Limited Areas

The islanders wanted independent authority to determine which areas within their own EEZs would remain closed to foreign fishing vessels, principally to protect local fishing industries (Doulman 1986a, 10). In round seven, Solomon Islands, for example, proposed closing off 85 percent of its EEZ and licensing only five U.S. seiners, and Papua New Guinea proposed closing significant areas of its EEZ and its archipelagic waters (Wolfe 1986b, 3). Later, even Australia suggested it wanted to close certain areas within its EEZ (Wolfe 1986d). Because the United States would pay more for access to larger ocean areas, this issue was related to fees. The issue of closed and limited-entry areas in Solomon Islands and Papua New Guinea remained unsettled at the end of the tenth round, and the United States engaged in bilateral negotiations with Australia, Papua New Guinea, Solomon Islands, and other concerned nations between that round and the final signing ceremony in early 1987 (Wolfe 1986d).

Catch Reporting

A contentious issue emerged at the sixth round of negotiations. Members of the U.S. delegation were concerned about what they perceived to be overly burdensome reporting requirements. Under the islanders' proposal, U.S. vessels would be subject to national reporting requirements within each zone (Wolfe 1986a, 2). The ATA argued that its data were confidential, and it resisted agreeing to reporting requirements. At the tenth round, in Tonga, it accepted the uniform South Pacific Commission standard reporting procedures, resolving the issue of reporting catch data (Doulman 1986d).

Licensing Fees

The greatest stumbling block in the negotiating process was the amount of the licensing fees, a subject first discussed at round seven in Canberra, Australia, in March 1986 (Wolfe 1986b, 1). At that meeting, the islanders named an annual price of $20 million for the first year, giving three reasons: (1) Tuna, the region's primary natural resource, is the "cornerstone" of regional economic development. (2) Foreign nations other than the United States offer competitive fees. (3) Creating one multilateral agreement with standard regional terms is a valuable convenience for U.S. fishing interests. The statement alluded to a fourth reason, the strategic issue of Soviet presence in the region, saying, "One power with whom we do not have the same traditional ties as with the United States" is competing for access to the region (Wolfe 1986b, attachment). (See Appendix 2 for excerpts of the islanders' statement.)

The United States responded at the eighth round, in Honolulu, in May 1986, by offering about $50,000 per vessel—a total of about $1.5 million annually in license fees for the 30 boats operating in the region—plus $4 million in increased economic assistance. (See Appendix 3 for excerpts from the U.S. statement on fees, Appendix 4 for the Secretary of State's message to the Pacific island delegations.) The Secretary of State said that the fees should be related to the economics of fishing, stressing that the U.S. government could not subsidize the fishing industry:

> We have consistently stated that access fees must be based on the economics of the fishery, and that our fishing boats cannot pay more for a license than it is worth commercially. From the United States point of view, it cannot provide a subsidy to the fishing industry to support its operations in the Pacific fishery. That is to say, Counselor Derwinski and Ambassador Wolfe cannot agree to any fee structure that requires the United States Government to participate in the funding. (Shultz 1986, 1)

The islanders answered that direct payment rather than other forms of economic aid was required and that the counteroffer was far too low: "Any arrangement which allowed boat owners such low payments would be completely unacceptable." (See Appendix 5 for excerpts from the FFA response of May 9, 1986, to the statement of May 8, 1986).

At the ninth round, in the Cook Islands, the United States offered $7.5 million in cash payments, including per-vessel licensing fees likely to total $1.5 million. The islanders had based their $20 million valuation on the price the Japanese, the Russians, and others had been willing to pay for access to certain zones in the Pacific, but they lowered their request first to $17.5 million annually, then to $33 million for a term of two years (Wolfe 1986c, 1).

Finally, after an intense week at the tenth round of negotiations in Nuku'alofa, Tonga, in October 1986, the United States reached agreement with the island nations on a price that amounts to about $60 million over a five-year period. Under the terms of the agreement, the U.S. government will pay the FFA $9 million annually in cash and will provide another $1 million per year for economic development projects in the region "formulated by the Forum Fisheries Agency" and administered through the U.S. Agency for International Development (USAID) office based in Fiji. This U.S. government commitment "is additional to existing USAID development assistance programmes already operational or planned in the South Pacific region." The ATA also agreed to pay annual license fees of $50,000 per vessel and guaranteed a minimum of 35 vessels (for a total of $1.75 million) for at least the first year of the agreement. If more than 35 ATA vessels enter the region during the first year, 10 additional licenses can be purchased for $60,000 per vessel. During subsequent years, up to 40 licenses can be purchased for $50,000 each, and 10 more for $60,000 each, but these prices will rise proportionately if the price of the tuna rises. Finally, the USTF will provide in-kind technical assistance worth $250,000 per year, bringing to at least $2 million the total contribution of the private tuna industry during the first year. (See the Agreed Minute in Appendix 6.)

The FFA will apportion the cash payments among the island nations. Each of the 16 nations will receive a minimum guaranteed amount; the remaining 85 percent will be distributed according to the fishing vessels' catch locations. (According to U.S. law, the development aid cannot go to Australia or New Zealand, because they are not deemed developing countries.) Also included in the final agreement is a stipulation that there will be a closed area in Solomon Islands not available for U.S. fishing (Doulman 1986d; Honolulu Star Bulletin 1986, A–12; Honolulu Advertiser 1986, D–1). The settlement implicitly recognizes the authority of the Pacific island nations over the tuna in their EEZs by virtue of the U.S. agreement to pay these amounts for fishing rights in the region. The first introductory paragraph of the 1987 treaty repeats language from Article 56 of the 1982 Law of the Sea Convention:

> ACKNOWLEDGING that in accordance with international law, coastal States have sovereign rights for the purposes of exploring and exploiting, conserving and managing the fisheries resources of their exclusive economic zones or fisheries zones . . .

This language—included in a treaty explicitly concerned with highly migratory species—appears to be a recognition that the highly migratory resources are included in the "sovereign" resources of the coastal nation when they are caught within the 200-mile zone of that coastal nation. A subsequent treaty provision in Annex I, Part 1 (3), seems designed to enable the United States to reserve its position on this matter, at least in a formal sense, but the White House press release announcing that the treaty had been completed reports this matter using ordinary language that acknowledges that the United States has agreed to pay the island nations for a resource that is recognized to be theirs:

> On October 20, 1986, negotiators from the U.S. and 16 Pacific island nations reached agreement on a regional fisheries treaty that will give American tuna vessels access to some 10 million square miles of rich fishing grounds in the South Pacific Ocean. The agreement provides just and fair compensation to the islands for the

resource, and offers the parties to the Treaty a substantial development assistance package that will continue the long tradition of close and productive relations between the U.S. and the island states. (White House 1986)

CONCLUSION

This treaty is a mutually satisfactory way of resolving the long-festering dispute between the United States and its neighbors in the Pacific. The new treaty acknowledges that coastal nations have jurisdiction over fishery resources while they are in their EEZs and that it is reasonable for distant-water fishing nations to pay for the fish they harvest at a fee level determined by market forces and political considerations. The recognition that the island nations have rights over the tuna in their 200-mile zones brings U.S. policies in line with the policies of the rest of the world. The treaty establishes a program of sharing the tuna resources that should benefit both the islands and the distant-water fishing nations.

NOTE

1. Copyright © 1987 by Jon M. Van Dyke and Carolyn Nicol.

REFERENCES

Agreed Minute

1986   Relating to financial arrangements for the proposed treaty on fisheries between certain Pacific island states and the United States of America. Signed by Philip Muller, director of the South Pacific Forum Fisheries Agency, and Edward E. Wolfe, representative of the U.S. government, October 20. U.S. Department of State. Washington, D.C.

Bugosh, Pamela

1984   The United States tuna policy as protest against the emerging customary law regarding highly migratory species. Center for Research and Advanced Study. University of Maine. Augusta, Maine. Mimeo. 29 pp.

Burke, William T.

1984   Highly migratory species in the new law of the ocean. Ocean Development and International Law 63:273–314.

1985   The Law of the Sea Convention and fishing practices of nonsignatories, with special reference to the United States.

In Consensus and confrontation: The United States and the Law of the Sea Convention. Edited by Jon M. Van Dyke. Law of the Sea Institute. Honolulu. pp. 314–337.

Coase, R. A.

1982   Statement of facts concerning F.F.V. [foreign fishing vessel] *Danica* presented to Rabaul District Court, Papua New Guinea, February 18. 3 pp.

Diro, Ted

1982   Election speech. Times. Port Moresby, Papua New Guinea. May 7. p. 12.

Doulman, David J.

1986a   Round six of the Pacific tuna treaty talks: Issues and developments. Pacific Islands Development Program. East-West Center. Honolulu. 13 pp.

1986b   Import prohibitions of tuna and tuna products. Pacific Islands Development Program. East-West Center. Honolulu. Mimeo. 1 p.

1986c   Interview, July 22. Pacific Islands Development Program. East-West Center. Honolulu.

1986d   Interview, October 29. Pacific Islands Development Program. East-West Center. Honolulu.

Felando, August

1982   Law of the Sea: The U.S. tuna industry perspective. Statement submitted to the House Committee on Merchant Marine and Fisheries, August 18. U.S. House of Representatives. Washington, D.C. 17 pp.

1986   Abstract accompanying "A perspective from the U.S. tuna seiner fleet: Has there been a change in policies by the U.S. government?" Paper presented at the 37th annual meeting of the tuna conference. National Marine Fisheries Service. Lake Arrowhead, California. Mimeo. 2 pp.

Fishermen's Protective Act

1967   U.S. Code, vol. 22, secs. 1971-79 (1976).

Fishery Conservation and Management Act

1976   Public Law 94–265, 90 Stat. 340 (1976), U.S. Code, vol. 16, secs. 1801–82 (1976 and Supp. IV 1982).

GATT (General Agreement on Tariffs and Trade) panel

1981　Report of General Agreement on Tariffs and Trade panel on United States prohibition of imports of tuna and tuna products from Canada. Restricted document L/5198. Geneva. 27 pp.

Gorton, Slade

1981　Statement on treaty with Canada on Pacific coast albacore tuna vessels and port privileges, July 20, 1981. U.S. Senate. 97th Cong., 1st sess. Congressional Record 127(13):16462.

Harrison, Craig S.

1985　Costs to the United States in fisheries by not joining the Law of the Sea Convention. In Consensus and confrontation: The United States and the Law of the Sea Convention. Edited by Jon M. Van Dyke. Law of the Sea Institute. Honolulu. pp. 352–368.

Herrick, Samuel F., and Steven J. Koplin

1984　U.S. tuna trade summary 1984. Administrative report no. SWR–85–6. National Marine Fisheries Service. Terminal Island, California. 36 pp.

Honolulu Advertiser

1986　Pacific tuna accord. October 22. p. D–1.

Honolulu Star-Bulletin

1986　U.S. to pay 16 Pacific islands $60 million for fishing rights. October 20. p. A–12.

Islands Business (Suva, Fiji)

1985　Fish talks speed up. November. pp. 26–27.

1986a　Tagging tuna shoals. January. p. 26.

1986b　Talking tuna and U.S. law. August. pp. 14–15.

Japan Whaling Association v. American Cetacean Society

1986　U.S. Supreme Court. The United States Law Week 54: 4929–37.

Joseph, James

1977　The management of highly migratory species: Some important concepts. Marine Policy 1(4)(October):275–288.

Kager, Levi

1982   Boarding officer report concerning F.F.V. [foreign fishing vessel] *Danica* presented to Rabaul District Court, Papua New Guinea, February 8. 2 pp.

Keith-Reid, Robert

1984   War on the tuna pirates. Islands Business. Suva, Fiji. August. pp. 10–12.

1986   Haggling over the price of fish. Islands Business. Suva, Fiji. August. pp. 10–12.

King, Joan

1986   *Priscilla M* freed after settlement. Sunday News. Guam. September 28. p. 3.

Kronmiller, Theodore G.

1981   Statement on the American tuna act, S.1564, at hearing before the National Ocean Policy Study of the Committee on Commerce, Science, and Transportation, December 8. U.S. Senate. 97th Cong., 1st sess. CIS 1982 S261–25 Serial no. 97–85. pp. 3–11.

1983   Exclusive economic zone of the United States: The question of sovereign rights relative to tuna. Statement before the U.S. Senate Committee on Foreign Relations, June 7. Reprinted in Ocean Science News 27(23)(June 13, 1983):2.

Laumaea, Susuve

1982a   We will not bow down: Message to America. Post-Courier. Port Moresby, Papua New Guinea. March 4. p. 1.

1982b   Open go for U.S. boats and *Danica* goes home—for K200,000. Post-Courier. Port Moresby, Papua New Guinea. March 12. p. 1.

Lebasi, Biga

1982   Letter to the editor. Times. Port Moresby, Papua New Guinea. March 3. p. 2.

McCloskey, Paul N., Jr.

1982   Statement to the House of Representatives on the Papua New Guinea crisis, March 11. U.S. House. 97th Cong., 2d sess. Congressional Record 128(3):3947–48.

National Union (Pohnpei, Federated States of Micronesia)
1985   U.S. tuna boat illegally fishing in FSM waters. June 30. p. 5.

Niugini News (Port Moresby, Papua New Guinea)
1982   PNG gov't signs deal with tuna fleet chief. March 19. p. 5.

Pacific Islands Monthly (Sydney, Australia)
1982   Seven-nation Nauru agreement shows its teeth in *Danica* seizure. May. pp. 51–53.

Percy, Charles
1981   Statement on treaty with Canada on Pacific coast albacore tuna vessels and port privileges, June 20. U.S. Senate. 97th Cong., 1st sess. Congressional Record 127(13):16462.

Persevero, P.
1982   Letter to the editor. Times. Port Moresby, Papua New Guinea. April 15. p. 2.

Post-Courier (Port Moresby, Papua New Guinea)
1982   We want *Danica*. March 15. p. 2.

Shultz, George P.
1986   Message to Pacific Islands delegations. Eighth round of negotiations. Honolulu. U.S. Department of State. Washington, D.C. May 8. 1 p.

Times (Port Moresby, Papua New Guinea)
1982   The Times Opinion. February 19. p. 32.

Trumbull, Robert
1982   Tuna war in south Pacific costing U.S. lots of good will. Honolulu Star-Bulletin and Advertiser. April 25, p. B–3.

Tsamenyi, B. Martin
1985   The South Pacific states, the U.S.A. and sovereignty over highly migratory species. Marine Policy 10(1):29–41.

U.S. Code, vol. 16
          The Fishery Conservation and Management Act of 1976 (FCMA). Public Law 94–265, 90 Stat. 340 (1976). U.S. Code, vol. 16, secs. 1801–02, 1803, 1822, and 1825 (1976 and Supp. 1986).

U.S. Code, vol. 22

Protection of vessels on the high seas and territorial waters of foreign nations. The Fishermen's Protective Act of 1967. U.S. Code, vol. 22, secs. 1971, 1975, and 1977 (1976 and Supp. 1986).

Van Dyke, Jon, and Susan Heftel

1981   Tuna management in the Pacific: An analysis of the South Pacific Forum Fisheries Agency. University of Hawaii Law Review 3(1):1–65.

Weicker, Lowell P., Jr.

1981   Statement on the American Tuna Protection Act, S.1564, at hearing before the U.S. Senate Committee on Commerce, Science, and Transportation, July 31. Congressional Record S9106–07.

White House

1986   Principal Deputy Press Secretary. Statement at Waukesha, Wisconsin, October 23.

Wolfe, Edward E.

1985   Delegation report at western Pacific regional tuna negotiations, Apia, Western Samoa, October 21–25. U.S. Department of State. Washington, D.C. Mimeo. 5 pp.

1986a   Delegation report at western Pacific regional tuna negotiations, Kona, Hawaii. January 27–February 1. U.S. Department of State. Washington, D.C. Mimeo. 4 pp.

1986b   Delegation report at western Pacific regional tuna negotiations, Canberra, Australia, March 20–26. U.S. Department of State. Washington, D.C. Mimeo. 6 pp.

1986c   Delegation report at western Pacific regional tuna negotiations, Rarotonga, Cook Islands, July 11–17. U.S. Department of State. Washington, D.C. Mimeo. 1 p.

1986d   Telephone interview, December 1. U.S. Department of State. Washington, D.C.

APPENDICES

Appendix 1. 1982 Law of the Sea Convention
Articles 56 and 64

Article 56—Rights, jurisdiction and duties of the coastal State in the exclusive economic zone

1. In the exclusive economic zone, the coastal State has:

(a) sovereign rights for the purpose of exploring and exploiting, conserving and managing the natural resources, whether living or non-living, of the waters superjacent to the sea-bed and of the sea-bed and its subsoil. . . .

(c) other rights and duties provided for in this Convention.

2. In exercising its rights and performing its duties under this Convention in the exclusive economic zone, the coastal State shall have due regard to the rights and duties of other States and shall act in a manner compatible with the provisions of this Convention.

Article 64—Highly migratory species

1. The coastal State and other States whose nationals fish in the region for the highly migratory species listed in Annex I shall cooperate directly or through appropriate international organizations with a view to ensuring conservation and promoting the objective of optimum utilization of such species throughout the region, both within and beyond the exclusive economic zone. In regions for which no appropriate international organization exists, the coastal State and other States whose nationals harvest these species in the region shall cooperate to establish such an organization and participate in its work.

2. The provisions of paragraph 1 apply in addition to the other provisions of the Part.

Appendix 2. Statement of Pacific island nations
on license fees

[Presented at western Pacific regional tuna negotiations, Canberra, Australia, March 1986]

. . . We have based our fee level on three main factors—the value of the resources to us, the interest of others in access to the resources, and the nature of the Agreement.

. . . We are all committed to developing tuna industries as a cornerstone of our plans for economic development. Some of us are already involved in the tuna industry, and our plans include possible cooperative ventures among us.

. . . Other foreign interests also seek access to these resources keenly and are coming to terms with the value and importance of these fish to us. We currently have arrangements for over 1,000 foreign tuna boats to fish in our waters, . . . land and process fish in our countries and employ our nationals on their vessels. There continue to be approaches for new arrangements from other foreign interests . . . [including] one power with whom we do not have the same traditional ties as with the United States.

. . . While we are fully committed to regional cooperation in fisheries, our governments have so far strongly preferred to enter into bilateral arrangements. It was a huge concession by our governments to agree to a multilateral arrangement with standard regional terms and conditions to meet your needs. Having a single regional arrangement subjects us to risks, costs and complexities and benefits your fishermen. It has no precedent.

. . . It is your fishermen and your government that will benefit from not having to make separate arrangements with each of us. Your fishermen will with a single license get stable access to the most productive tuna fishing grounds in the world. . . .

. . . The tuna are our major natural resource. They support our plans for social and economic development. We will not compromise on their value. The fee is US$20 million for the first year.

Appendix 3. U.S. statement on fees

[Presented at the eighth round of negotiations between the FFA members and the United States on a multilateral fisheries treaty, May 8, 1986, Honolulu, Hawaii]

The most difficult and important issues in our negotiations are those of closed areas and license fees, which in our view are inextricably linked. Regarding license fees, . . . from our point of view, we should pursue a two-tiered approach. . . . A fundamental reason for this is that the U.S. government cannot contribute funds towards a fishing fee as part of this agreement. Nonetheless, we are willing to commit the U.S. government to a financial role outside the context of the fisheries agreement. These points are underscored in a message from Secretary of State Shultz which we have just been authorized to present to you. . . . [Appendix 4].

Secretary Shultz has indicated that he will consider an expanded assistance program during the life of the agreement for those states which participate in the treaty. Of course, assistance can take many forms. It is our strong preference that appropriate developmental assistance projects form the bulk of our assistance. However, we would welcome further consultation with you on the precise form you would like this assistance to take.

The amount of money which we are prepared to offer under this two-tiered approach would of necessity also be separated into two components. The amount of expanded assistance which the U.S. government is prepared to offer to the Pacific island states is 4 million dollars annually . . . in addition to the expanded fisheries assistance program. . . .

The amount designated as license fees pursuant to the agreement will be strictly an industry payment. If the closed areas can be satisfactorily resolved, we can offer on behalf of our industry a fee of $50,000 per vessel. There are approximately 30 vessels operating in the region . . . [so] fishing fees pursuant to the agreement would be 1.5 million dollars. We would be willing to pursue with our industry the idea that the amount of the license fee would be tied to the price of fish. . . .

Appendix 4. Message of U.S. Secretary of State George P. Shultz to the Pacific island delegations

[Presented at the eighth round of negotiations between the FFA members and the United States on a multilateral fisheries treaty, May 8, 1986, Honolulu, Hawaii]

I want to emphasize the importance the United States places on maintaining and building on the excellent relations it enjoys with the nations of the Pacific. In that regard, I place a high priority on bringing these negotiations in which you are engaged to an early and successful conclusion.

. . . I want specifically to describe our views on the fee issue, as it involves fundamental principles of President Reagan's administration. We have consistently stated that access fees must be based on the economics of the fishery, and that our fishing boats cannot pay more for a license than it is worth commercially. From the United States point of view, it cannot provide a subsidy to the fishing industry to support its operations in the Pacific fishery. That is to say, Counselor Derwinski and Ambassador Wolfe cannot agree to any fee structure that requires the United States government to participate in the funding.

However, we are prepared to consider an expanded assistance program during the life of the agreement for those states which participate in the treaty. . . .

. . . I think we all recognize that the successful negotiation of a multiyear regional fisheries treaty will bring direct benefit to all of the countries involved, both yours and mine.

Appendix 5. Response of the Forum Fisheries Agency of May 9, 1986, to the U.S. Statement of May 8, 1986

[Presented at the eighth round of negotiations between the FFA members and the United States on a multilateral fisheries treaty, May 9, 1986, Honolulu, Hawaii]

. . . We have no objection in principle to the fee arrangement having two components. Our final attitude will however depend on the form and level of the component to be provided by the U.S. government as well as the total level of the package.

. . . [W]e are very disappointed at the overall level of fees. . . . It falls far short of what we had sought under a regional treaty, and we hope you will be in a position to bring forward a realistic fee proposal as the fee discussions proceed.

Within the overall fee level you propose, the industry contribution is completely inadequate, on any criteria. We expect the industry to pay substantially more than the amount you have suggested. Whatever the structure of fees, any arrangement which allowed boat owners such low payments would be completely unacceptable to us.

. . . We continue to prefer a lump sum basis for all payments. We require that the fees should be direct financial payments.

. . . [U.S. Government assistance would be unacceptable] in light of a resolution adopted by our Heads of Government at the South Pacific Forum in Rotorua in 1982, rejecting any tying of aid to access fees.

Appendix 6. Agreed Minute

The Representatives of the Governments of Pacific Island States and the Government of the United States of America meeting in Nuku'alofa, Kingdom of Tonga, consider it desirable to record the points set forth below relating to financial arrangements for the proposed treaty on fisheries between certain Pacific Island States and the United States of America.

(1) The Term would be for five years.

(2) $10 million annual U.S. Government cash grant, of which $1 million would be used to fund economic development projects formulated by the Forum Fisheries Agency. It is understood that the U.S.

Government commitment is additional to existing USAID development assistance programmes already operational or planned in the South Pacific region. The $10 million annual Government cash grant will be administered in accordance with an Agreement between representatives of the United States Government and the South Pacific Forum Fisheries Agency.

(3) Annual industry payment for licence fees as follows:

(a) for the first year, a guaranteed lump sum of $1.75 million for 35 vessels, with the next 5 licences to be made available for the same pro rata fee as the first 35 licences, and an additional 10 licences to be made available at $60,000 per vessel;

(b) for subsequent years, 40 vessel licences calculated on the same fee basis as the first 40 vessel licences in paragraph (a) and indexed to the price of fish, with 10 additional licences to be made available at $60,000 per vessel and indexed to the price of fish, provided that in no case shall the level of the per vessel licence fees fall below the level of vessel licence fees for the first year.

(4) Technical assistance valued at $250,000 annually by the U.S. tuna industry, in response to requests co-ordinated through the Forum Fisheries Agency. This would include assistance at the technical level if requested. The value of this total financial package over five years is estimated as likely to exceed $60 million.

Done at Nuku'alofa, Kingdom of Tonga, on the 20th day of October, 1986.

Philip Muller
Director of the South Pacific
Forum Fisheries Agency,
as Representative of the
South Pacific Forum Fisheries
Agency Member States

Edward E. Wolfe
Representative of the
Government of the United
States of America

8.
Development and Expansion
of the Tuna Purse Seine Fishery

David J. Doulman

INTRODUCTION

Purse seining is the most capital-intensive, technologically sophisticated, and efficient means of catching tuna.[1] It is the principal method used for harvesting tuna for the world's canned-tuna market.[2] Because of its efficiency, purse seining has largely displaced pole-and-line fishing, which—along with some longline fishing principally for albacore tuna (*Thunnus alalunga*)—has traditionally supplied the canning industry.[3]

The purse seine fishery in the Pacific islands is the region's newest tuna fishery. Purse seine catches account for more than half the region's commercial tuna catch each year. However, longline catches outrank purse seine catches in value because the purse seine fishery targets on the more abundant but commercially less valuable surface-swimming tuna species—skipjack (*Katsuwonus pelamis*) and juvenile yellowfin (*Thunnus albacares*).

Tuna purse seining began in the Pacific islands region in the early 1970s and escalated during the 1980s with the entrance of the U.S. fleet and the expansion of Asian fleets. Its rise has paralleled the decline of the region's longline and pole-and-line fisheries that began in the 1970s and continues today.

This paper reviews the development and examines the status of the purse seine fishery in the Pacific islands. It first supplies background information about the development of the purse seine fishing technique in the United States and its adoption and application in the Pacific islands region by the Japanese. Succeeding sections review the expansion of the fishery, identify the characteristics of single seiners and group seiners, compare the operations of Japanese and U.S. fleets, outline the arrangements by which foreign fleets gain access to the fishery, discuss fleets based in the region and current and potential transshipping arrangements, and analyze catches of U.S. and Japanese fleets from 1980 to 1986. The final section speculates on likely directions in the purse seine fishery.

BACKGROUND

U.S. Purse Seine Technology

Purse seining for tuna was pioneered by fishermen from San Pedro and San Diego on the U.S. west coast. The first U.S. seiner designed for tuna fishing was commissioned by the Van Camp fishing and canning corporation in 1916, but U.S. fishermen did not start purse seining in the eastern tropical Pacific until 1925 (Green et al. 1971, 182–184).[4]

Low-priced tuna imports from Japan in the 1950s pushed U.S. tuna fishermen into improving their efficiency in order to remain competitive. They started converting their pole-and-line vessels to purse seiners; some of them commissioned new seiners, in many cases with financial backing and logistical support from U.S. canners. These moves, coupled with the development of nylon fishing nets and the installation of power blocks on U.S. vessels, revolutionized purse seining, making it the most efficient tuna fishing method known. By improving and fine-tuning gear and vessel designs and building larger and faster vessels, the U.S. tuna industry maintained its superiority in purse seine fishing. By 1980 the U.S. purse seine fleet consisted of about 150 vessels. Figure 1 shows a schematic of purse seining.

Japanese Fisheries Expansion

A year after the 1952 lifting of the MacArthur Line that had restricted its postwar fishing efforts, Japan enacted legislation to rebuild and expand its fishing industry. Tuna-fishing enterprises turned their attention to new technologies and methods for improving efficiency. As part of its fisheries revitalization program, Japan officially sponsored initiatives to improve the operations of its tuna fleet.[5] For purse seining, however, both government and industry saw merit in using U.S. vessel designs and technology rather than trying to develop and promote their own new technology. The Japanese wanted to adapt U.S. technology for use in trial programs aimed at developing new offshore tuna fishing grounds. They were particularly interested in trying to develop a tuna purse seine fishery in tropical waters south of Japan.[6]

In the late 1950s and 1960s Japan built several purse seiners of U.S. design incorporating U.S. technology. These vessels were built primarily for commercial operations; some were also used for trial fishing.[7]

The U.S.-style seiners were first deployed in Japan's coastal waters, but between 1964 and 1974 up to six seiners fished around Micronesia

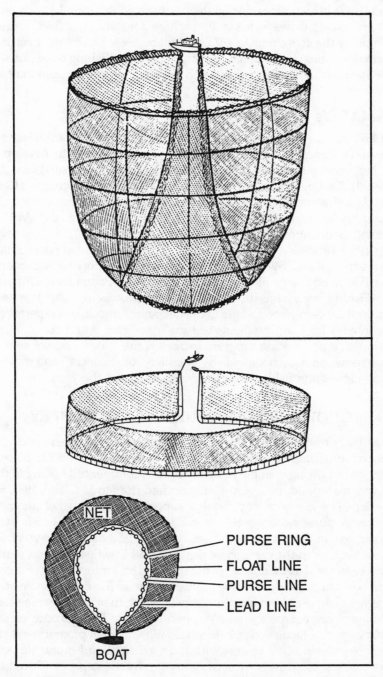

Figure 1. Illustration of setting and pursing a purse seine net

and Papua New Guinea during the northern winter months each year. Their annual catches were small, reaching a maximum of 3,392 tonnes caught by the six seiners in 1974 (Watanabe 1983, 5). The poor catches were due more to unfamiliarity with purse seine fishing conditions in the Pacific islands region than with inability to use the technique effectively.

JAMARC's Research Program

In 1971 the government of Japan established the Japan Marine Fishery Resource Research Center (JAMARC) and charged it to develop a distant-water purse seine tuna fishery (*kaigai maki-ami gyogyo*) (Watanabe 1983, 1). Exploratory fishing was undertaken off the west coasts of Central America and Africa in order to master the U.S. purse seine technique under tropical fishing conditions. Achieving this goal, JAMARC shifted its attention back to the Asia-Pacific region in 1973. Trial seining in the Timor Sea off the north coast of Australia yielded poor results. Between 1974 and 1976, JAMARC conducted exploratory fishing around the Federated States of Micronesia, Palau, and Papua New Guinea.[8]

Fishing conditions in this area were promising. The Japanese adapted the seining technique to the deeper thermocline and perfected techniques for fishing schools of tuna associated with floating logs.[9] JAMARC's purse seine program increased the success rate of setting on free-swimming schools of fish ("splashers" or "foamers") and showed that purse seine fishing was feasible throughout the year.

GROWTH OF THE PURSE SEINE FISHERY

JAMARC's results paved the way for commercial development and later internationalization of the purse seine fishery in the Pacific islands region. In 1976 eight single Japanese seiners commenced fishing in the region year-round; by 1982 the number had risen to 33 (Habib 1984, 6).

Japanese government policy has kept to 31 the number of commercial single purse seine vessels licensed to operate in the Pacific islands.[10] Expansion in the purse seine fleet between 1978 and 1982 was offset by reductions in the capacity of the pole-and-line fleet, caused partly by a vessel replacement policy that required an entrepreneur or a corporation wanting a purse seine license to acquire five existing pole-and-line vessel licenses from other fishermen, which then were not renewed. This replacement policy was stopped in 1982. The purpose of the policy—part of Japan's overall fisheries rationalization program—was to replace pole-and-line vessels with more efficient and more competitive purse seine vessels without losing productive capacity (Doulman and Wright 1983, 55–56).

The initial success and profitability of Japan's single purse seine fleet operations in the Pacific islands provoked interest from Japan's group seine tuna fleet and from the purse seine fleets of other distant-water fishing nations (DWFNs), encouraging them to join the fishery.

Japan's group seine tuna fleet first operated in the islands region in Federated States of Micronesia in 1980. These groups of vessels, which normally operated in Japan's coastal waters, decided to extend their operations southward during the northern winter. Four group seiners with their support vessels operated in the fishery in 1980; in 1981 they moved southward into Papua New Guinea's exclusive economic zone (EEZ). The number of Japanese group seiners operating in the region on a distant-water basis has since increased to seven, though nine groups are eligible to fish (Table 1).

Throughout the late 1970s U.S. purse seine fishermen monitored Japanese progress in the western Pacific purse seine fishery. The

Table 1. Purse seine fleets registered to operate in the EEZs of South Pacific Forum member countries 1985

	Single	% of single purse seiners[a]	Group	% of group purse seiners	Total purse seine fleets
Cayman Islands	3	1			3
Indonesia	1	1			1
Japan	33	22	9	82	42
Korea	9	6			9
Mexico	2	1			2
New Zealand	2	1			2
Panama	6	4			6
Philippines	13	9			13
Taiwan	4	3	2	18	6
USA	60	41			60
USSR	13	9			13
Vanuatu[b]	1	1			1
Total	147	100	11	100	158

Source: Clark 1986, 157.

[a]Percentages may not add to 100 due to rounding.

[b]Vessel registered in Vanuatu but owned by Ecuadorian interests. The vessel normally operates in the eastern Pacific (Doulman and Kearney 1986).

Japanese proved that purse seining could succeed in the region despite the poor catch rates of U.S. seiners during exploratory trips in the waters of Micronesia and Papua New Guinea in the 1970s.[11] The reported average daily catch rates of the U.S. vessels had ranged from 3.8 tonnes per day in 1976 to 11.9 tonnes in 1978 (Papua New Guinea 1982, 23).

The overcapitalization of the U.S. fleet, the low and declining catch rates, and deteriorating relations in the 1970s with Latin American coastal states over access to their EEZs had prompted U.S. fishermen to move into the Pacific islands region. This was an attractive option because research by the South Pacific Commission had found large skipjack tuna resources in the region—resources that seemed capable of sustaining higher levels of exploitation than were being applied. U.S. fishermen believed that if they could adapt their fishing techniques as the Japanese had, they could enhance their catch rates and their financial returns.[12]

For many U.S. fishermen an important consideration in the decision to move into the western Pacific was the onset of the 1980–81 El Niño. Partly to escape the effects of El Niño in the eastern tropical Pacific, they started commercial fishing in Micronesia (Federated States of Micronesia, Marshall Islands, and Palau) in 1980 under terms of an agreement concluded between the American Tunaboat Association (ATA) and the three Micronesian states.

Purse seine fleets from other DWFNs followed the lead of Japan and the United States and began to deploy vessels in the region. They too had observed the pioneering efforts of the Japanese in developing the fishery and realized that the fishing grounds in the Pacific islands were among the most productive in the world.

The growth of the purse seine fishery in the Pacific islands has been spectacular. More than 120 single and group seiners operated in the region in 1984 (Doulman 1986a, 11). The number of single seiners operating declined in 1985 and 1986, primarily because fewer U.S. vessels sought fisheries access there. The record number of seiners in 1984 is unlikely to be exceeded.

PURSE SEINE FLEETS

In 1985 there were 158 single and group purse seiners from 11 countries listed on the Regional Register of Fishing Vessels maintained by the Forum Fisheries Agency (FFA) (Table 1). This is the maximum number that could legally operate in the EEZs of FFA member countries; in any year the number of seiners actually fishing is likely to be well below the number registered.

Single Seiners

FFA's vessel register lists 147 single purse seiners. The majority of the seiners are owned and controlled by Japanese and U.S. individuals and corporations. Including the U.S.-owned vessels of Cayman Island and Panama registry, 102 seiners (69 percent) are concentrated in Japanese and U.S. hands (Table 1). A number of other seiners—particularly those from Korea, the Philippines, and Taiwan—also have direct or indirect ties with U.S. fishing and processing interests.

Until 1985 all but one of the Japanese single seiners were 499-GRT (gross registered tonne) vessels (Table 2).[13] This homogeneity is largely due to JAMARC'S success with vessels of this class during exploratory fishing and to Japanese government licensing policy. However, three 499-GRT vessels have been replaced by 750-GRT vessels, one in 1985 and two in 1986. The trend for replacing vessels in the Japanese fleet seems to be toward larger-class vessels.[14] Thus, while Japanese policy restricts the number of single purse seine licenses, the fleet's capacity is increasing.

Table 2. Gross registered tonnage of single purse seine fleets registered to operate in the EEZs of South Pacific Forum member countries 1985

	Gross registered tonnage				
	< 500	500–1,000	1,000–2,000	> 2,000	Total
Cayman Islands			2	1	3
Indonesia		1			1
Japan	32	1			33
Korea		7		2	9
Mexico			2		2
New Zealand	2				2
Panama		3	2	1	6
Philippines	5	8			13
Taiwan	1	2	1		4
USA		15	34	11	60
USSR		5		8	13
Vanuatu		1			1
Total	40	43	41	23	147
Percentage of total	27	29	28	16	100

Source: Clark 1986, 157.

All U.S. seiners listed on FFA's vessel register (and most foreign flag vessels belonging to U.S. corporations) are larger than 500 GRT. Indeed, 75 percent of the U.S. vessels registered are of the superseiner class— over 1,000 GRT.[15] Figure 2 shows the typical design of a modern U.S. purse seine vessel, complete with helicopter.

The largest purse seiner (3,000 GRT) operating in the islands region is under Russian flag.

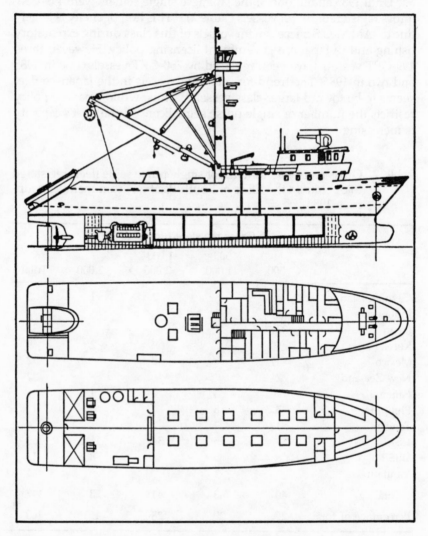

Figure 2. Design of a typical U.S. purse seine vessel

Nine Korean single purse seiners were listed on the FFA register in 1985 (Table 1). Seven of these were in the 500- to 1,000-GRT range; two were over 2,000 GRT (Table 2). However, according to a 1987 industry report, the Korean purse seine fleet now consists of 18 vessels.[16] These vessels are owned by at least seven corporations; one corporation (Dongwong Industries Co., Ltd.) owns seven of them. Of the active fleet, 12 seiners are ex-U.S. vessels, two are ex-French, and two are new U.S. fleet vessels that have never operated under the U.S. flag.

Franklin (1982) and Habib (1984) have noted that details of the Korean fleet's operations in the Pacific islands are scarce. However, the fleet transships tuna within the region at Guam and Tinian. It also has informal and formal ties with U.S. processors—in particular, Star-Kist—for the delivery of its catch.

In 1981 the Korean government set a policy limiting the number of purse seine vessel licenses, but that policy has reportedly been discontinued. Furthermore, the government proposed to reduce longline fishing effort in the region and to replace it with purse seine effort (Habib 1984, 16).[17]

The Taiwanese purse seine fleet consists of seven single and three group seiners. However, the single purse seine fleet is being increased; at least two, and possibly six, additional vessels are to be built in Taiwan in 1987. The single purse seine fleet is a new fleet: five of the seiners have been constructed in Taiwan since 1985—one in 1985, one in 1986, and three in 1987.[18] The remaining two single seiners had earlier been U.S. and Indonesian flag vessels.

In 1985 Taiwan had four single and two group seiners listed on FFA's register (Table 1). With the projected increase in the size of the Taiwanese fleet, it is likely to become more prominent in the islands region. Like the Korean fleet, the Taiwanese fleet has close ties with U.S. processors.

The size distribution of purse seine vessels listed on FFA's register in 1985 showed 27 percent of single seiners under 500 GRT, 29 percent between 500 and 1,000 GRT, 28 percent between 1,000 and 2,000 GRT, and 16 percent greater than 2,000 GRT (Table 2).

The total fleet of single purse seiners in the Pacific islands region is characterized by a "core" fleet of vessels that operate throughout the year. This core fleet consists of the entire Japanese fleet and some of the vessels from Cayman Islands, Korea, Panama, and Taiwan. In addition, a highly mobile "floating" fleet operates within the region and outside it. U.S. vessels are the major group in the floating fleet. Since 1981 many U.S. seiners have operated in both the eastern and western Pacific each year. The time these vessels spend fishing in the western Pacific is largely determined by fishing conditions in the eastern Pacific and the time it takes to harvest the annual quota set by the Inter-

American Tropical Tuna Commission (IATTC). Because the number of months needed to harvest the IATTC quota has progressively fallen, U.S. vessels have come to depend more and more on the central and western Pacific as an area of operation. The number of U.S. vessels operating in the eastern tropical Pacific tuna fishery declined from 125 vessels in 1979 to 49 vessels in 1985—a fall of 61 percent (USITC 1986, 160).

The core group of seiners fishing in the Pacific islands confines its operations to the EEZs of the Federated States of Micronesia, Kiribati, Palau, and Papua New Guinea and to the high-seas pockets between these zones. The Japanese fleet, the most important fleet in the core group, has purse seine access agreements with the Federated States of Micronesia, Palau, and Papua New Guinea (Table 3). It targets principally on the EEZs of the Federated States of Micronesia and Papua New Guinea. Japan's Fisheries Agency restricts purse seine fleet operations west of 180°, largely to prevent problems of interaction between purse seine vessels and pole-and-line vessels and between purse seine gear and longline gear. For this reason, Japan has no purse seine access agreement with Kiribati (Table 3).[19]

Table 3. Purse seine fishing access agreements in force 1985

	Federated States of Micronesia	Kiribati	Palau	Papua New Guinea
Purse seiners				
Single	x	x	x	x
Group	x			x
DWFNs				
Japan	x		x	x
Korea	x	x		
Mexico	x			
Taiwan	x			x
USSR		x		

Source: Doulman 1986b, 19–20.

The geographic distribution of fishing effort within the Pacific islands region by the floating fleet is more diverse than that of the core fleet. The floating part of the fleet fishes not only in the EEZs of the Federated States of Micronesia, Palau, and Papua New Guinea but also around Kiribati, the Marshall Islands, Nauru, and elsewhere.

Group Seiners

The operations of the group purse seine fleet in the Pacific islands are less important than those of the single purse seine fleet because their operations are seasonal and they deploy fewer vessels.[20] Eleven group seiners were listed on FFA's vessel register in 1985 (Table 1). Nine of these groups were Japanese; the other two were Taiwanese.

Group purse seining uses three or four vessels—usually a seiner of 116 GRT (Japanese) or 160 GRT (Taiwanese), one or two fish carrier vessels of about 325 GRT, and an anchor vessel of 45 GRT (Doulman and Wright 1983, 54). The Japanese groups usually use four vessels, the Taiwanese groups three.

On a distant-water basis, up to ten groups operate in two Pacific island countries, the Federated States of Micronesia and Papua New Guinea (Table 3). The remaining Japanese group seiner, which is larger than the others (180 GRT), operates throughout the year in Solomon Islands. It is owned by the Taiyo Fishery Corporation, the Japanese joint-venture partner in Solomon Islands' domestic tuna fishery.[21]

COMPARISON OF JAPANESE AND U.S. FLEETS

There are several striking differences between the Japanese and U.S. purse seine fleets operating in the Pacific islands region. Some of these were pointed out in the preceding section; others are listed in Table 4.

Table 4. Characteristics of Japanese and U.S. purse seine fleets operating in the Pacific islands region

	Japan		United States
	Single	Group	Single
Vessel size (GRT)	499	116 (seiners) 325 (carrier) 45 (anchor)	Mainly >1,000
Helicopter	No	No	Yes
Vessel ownership	Corporate	Corporate	Individual
Crew	18–24	23 (seiner) 15 (carrier) 7 (anchor)	20–22
No. of trips per year	5–7	1	Variable
Daily catch rate (tonnes)	30–35	17–20	40–45

Beyond the differences in vessel size in the two fleets, there are differences in mode of operation. U.S. vessels normally use a "spotter" aircraft during fishing operations; Japanese seiners do not. Japanese vessels tend to fish as a fleet and exchange information on the fishing grounds; U.S. vessels usually operate individually and competitively. Because Japanese skippers exchange information, they can coordinate their fishing without expensive aircraft support. But U.S. purse seine operations need spotter aircraft for locating tuna schools.[22]

In keeping with their mode of operation, U.S. vessels are commonly owned by individuals or by small family companies. In many instances—particularly before the 1981 onset of turmoil in the world tuna industry—U.S. vessel owners had close ties with U.S. canners, who often provided (1) bank security for the purchase of vessels (either outright loan guarantees or contracts to purchase their catches), (2) trip advances, or (3) guaranteed delivery prices for fish (forward contracts) in return for a commitment from vessel owners to deliver their catches to the company providing the backing. However, since the 1981 restructuring of the tuna industry the relationship between most U.S. vessel owners and U.S. canners has been largely severed. Moreover, in February 1985, 21 of the most prominent U.S. vessel owners jointly instituted legal proceedings against the three principal U.S. processors—Star-Kist, Ralston Purina, and Castle and Cooke (U.S. District Court 1985). The action seeks damages and injunctive relief totaling $1.3 billion from the processors under the U.S. Antitrust Act. The action is continuing. According to U.S. industry sources, the U.S. processors are vigorously opposing the action and are attempting to divide the vessel owners. The owners, however, are standing firm.

In contrast to U.S. practice, Japanese seiners tend to be owned by small- to medium-sized corporations, which often have more than one vessel. Moreover, the Japanese vessels are relatively new: 20 vessels (63 percent of the fleet) have been launched since 1981 (Habib 1984, 77). By comparison, just a few U.S. flag vessels have been launched since 1981. Many of the U.S. seiners built in the early 1980s were not delivered to their owners. Some were sold abroad; others remain tied up awaiting sale.

Japanese single seiners are crewed by 18 to 24 persons, group seiners and support vessels by about 45 persons, whereas U.S. seiners are crewed by 20 to 22 persons (Table 4). Japanese vessels are crewed by Japanese nationals only, but U.S. seiners employ a range of nationals, including islanders from American Samoa and Micronesia. Because it is expensive to employ Japanese and U.S. crews, U.S. vessel owners have sought to employ skilled fishermen from low-wage countries—primarily in Latin America—to reduce or contain costs. In many cases only the skipper and the aircraft pilot on U.S. vessels are U.S. citizens. If more seiners start to base their operations in the Pacific islands region, additional islanders will probably be employed, but their absolute number will be small.

Under normal fishing conditions in the Pacific islands, Japanese single seiners count on making five to seven fishing trips a year from their base port of Shimizu in Japan (Table 4). If fishing conditions are better than normal, up to nine trips per year can be made. Group seiners make only one trip lasting about four months (February to May) each year, then return to their traditional fishing grounds nearer Japan.

The number of fishing trips made each year by U.S. vessels in the islands region varies. Depending on a range of circumstances— including prevailing fishing conditions in the Pacific islands and in the eastern Pacific and arrangements for financial backing and delivery or transshipment of the harvest—seiners make up to six fishing trips per year (Table 4). Most vessels make only two or three trips.

Despite the difference in the size of Japanese and U.S. vessels, average daily catch rates for ther fleets do not vary much. The catch rate of Japanese single seiners averages 30 to 35 tonnes a day, while the larger U.S. vessels average 40 to 45 tonnes a day (Table 4). Group purse seiners catch about 17 to 20 tonnes of fish per day while operating in the region.

FISHERIES ACCESS

Pacific island countries license DWFN purse seine fleets under negotiated access agreements. In return for fishing rights, the fleets pay prescribed fees and agree to abide by certain terms and conditions. Some DWFNs also offer aid to island countries in an effort to secure more favorable access terms. However, FFA countries have agreed that fisheries-related aid (training, fishing supplies, etc.) and fisheries access should be treated as separate issues (SPEC 1984, 36), maintaining that access agreements should be based strictly on commercial considerations. License fees for purse seine vessels are normally related to the expected value of the catch within a certain time period. Because tuna prices fluctuate and fees are tied to monthly average prices, license fees vary by the month. Conceptually, this approach to determining fees is equitable for both island countries and DWFNs.[23]

Table 3 shows the DWFNs that had purse seine access agreements with Pacific island countries in 1985. The Federated States of Micronesia had agreements with four DWFNs (Japan, Korea, Mexico, and Taiwan), Kiribati with two (Korea and USSR), Palau with one (Japan), and Papua New Guinea with two (Japan and Taiwan). Because of their size and productivity, the EEZs of the Federated States of Micronesia and Papua New Guinea are the most important ones for purse seine fishing in the region.

All purse seine agreements in the region are concluded with island governments or, in the case of the Federated States of Micronesia, with a semigovernmental body, the Micronesian Maritime Authority. However, except for Korea, the DWFN parties to these agreements are industry groups (Table 5). Japan's purse seine agreements are concluded with two tuna fishing associations, Mexico's with a public corporation, Taiwan's with a private corporation, and the Soviet Union's with a public corporation.[24]

Although there have been no access agreements in the region for the U.S. fleet since December 1984, the Federated States of Micronesia and Papua New Guinea did not deny access to U.S. vessels.[25] Instead, they licensed them under the terms and conditions of other DWFN agreements. For example, Papua New Guinea licensed U.S. seiners and seiners from the Philippines and Korea under its agreement with Japan. In effect, these vessels were treated like Japanese flag vessels, paying the same fees and conforming to the same terms and conditions of access as the Japanese fleet did.[26]

Table 5. DWFN purse seine signatories to access agreements with Pacific island governments

	Classification	Organization
Japan	industry	tuna fishing associations
Korea	government	government
Mexico	industry	public corporation
Taiwan	industry	private corporation
USSR	industry	public corporation

Source: DWFN access agreements with Pacific island countries.

FLEETS BASED WITHIN THE REGION AND TUNA TRANSSHIPMENT

An estimated 29 purse seine vessels were based at ports or registered in the Pacific islands region in 1985 (Doulman and Kearney 1986). These vessels were based in American Samoa (10), Fiji (1), Guam (11), Nauru (2), Papua New Guinea (1), Solomon Islands (1), and Vanuatu (3). The owners' principal purpose in basing their seiners in the region is to gain financial benefits from being near the region's main fishing grounds

(Guam, Papua New Guinea, and Solomon Islands) or its processing facilities (American Samoa and Fiji). In recent years, harvests of purse seine vessels based in the region have been estimated at 140,000 to 160,000 tonnes of tuna annually (Doulman and Kearney 1986).

Most of the seiners based in the islands region are U.S. flag vessels or U.S. corporate-controlled vessels of flags of convenience. No Japanese single seiners base in the Pacific islands because the Japanese government officially discourages the use of ports outside Japan. Through this policy the government seeks (1) to control and monitor the operations of Japan's purse seine fleet, (2) to ensure that the fleet's catches are counted as exports of Japan, (3) to ensure that the benefits of fleet servicing accrue to Japanese industry, and (4) to try to limit supplies of tuna coming onto the Japanese market. The Japanese government also discourages Japanese seiners from transshipping their catches at foreign ports for similar reasons. However, despite government objections, some Japanese purse seiners started transshipping their catches at Tinian (Northern Marianas) in 1984.

Japanese policy on basing seiners in the region and transshipping catches there puts it at odds with some Pacific island governments that encourage transshipping. The Japanese purse seine fleet accepts the transshipment controls imposed by the Japanese government but, given the option, the fleet would probably seek to transship at least some of its catches in the region.

For the U.S. fleet and U.S.-controlled vessels, the advantages of basing vessels and transshipping harvests in the region have been demonstrated by the U.S.-owned fleet based in Guam.[27] This fleet, consisting initially of 8 corporately owned vessels and now of 11 vessels, moved to the western Pacific from the U.S. west coast in 1981. The fleet has made consistently good catches; one vessel took a record 7,000 tonnes of tuna in 1984. As a result of the fleet's move to Guam, unproductive time spent traveling or tied up in port has been minimized. Its operations are closely coordinated and controlled so that catches can be quickly transshipped to reefer vessels for transportation to canneries in American Samoa and elsewhere as necessary.[28]

Tuna transshipment from purse seine vessels regularly takes place at three locations in the region: Guam (Agana), Papua New Guinea (Rabaul), and the Northern Marianas (Tinian).[29] Transshipment of fish is accomplished by vessel-to-vessel transfer. Because these operations do not require shore-based investment, they are considered inherently temporary. In 1984 an estimated 80,000 tonnes of tuna was transshipped at these three locations (Doulman and Kearney 1986).[30]

Two Pacific island countries—the Federated States of Micronesia and Papua New Guinea—are contemplating investing in permanent transshipment facilities. Hawaii has also shown a desire to become a transshipment port, but although it has excellent servicing facilities, it is too far from the fishing grounds to have a geographic advantage. The future of permanent facilities in the region will depend on (1) the extent to which countries interpret and enforce Article 62 of the Law of the Sea Convention relating to the transshipment of catches at coastal states' ports, (2) the capacity of the countries to manage the facilities, and (3) the size of purse seine fleets operating in the region (Doulman and Kearney 1986).

In efforts to increase their participation in the purse seine tuna fishery in the region, some island governments and domestic investors have moved to purchase seiners or are considering doing so. Solomon Islands is having two 499-GRT seiners constructed in Australia; Kiribati and Papua New Guinea are investigating investment in vessels. Other island countries have expressed a similar desire from time to time, but not all of them can logistically support purse seine fleets because of their lack of such natural endowments as deepwater ports and such necessary service infrastructure as fuel storage facilities.

CATCHES

Tuna landed by purse seine fleets in the Pacific islands region is processed primarily at canneries in Japan, Taiwan, Thailand, the Philippines, and the United States (American Samoa and Puerto Rico). The canneries in Fiji and Solomon Islands do not generally process purse-seine-caught fish, but Fiji may be required to do so in future because it has experienced difficulties in securing sufficient quantities of longline- and pole-and-line-caught tuna to keep its cannery operating at capacity.

Because the value of purse seine tuna catches in the Pacific islands is a function of the quantity of fish harvested and prevailing market prices, the value of annual catches fluctuates widely. The value of purse seine catches in 1984 was estimated at $236 million (Doulman 1986b, 14). Of this total, the Japanese fleet accounted for 37 percent, the U.S. fleet (including corporate-owned vessels of flags of convenience) for 56 percent, and other fleets for 7 percent.

Japanese Fleet

Between 1980 and 1986 the Japanese single and group purse seine fleets harvested a total of 794,879 tonnes of tuna in the Pacific islands region (Table 6). Of this amount an estimated 590,000 tonnes was skipjack, 198,000 tonnes yellowfin, and 7,000 tonnes other species. Ninety percent of the total catch (716,550 tonnes) between 1980 and 1986 was taken by single purse seiners—a proportion relatively constant from year to year.

Table 6. Japanese single and group purse seine fleet catches in the Pacific islands region by species 1980–1986

	Tonnes			Percentage of total purse seine fleet catch	Single purse seine fleet catches as a percentage of the total catch
	Single	Group	Total		
1980	39,741	3,303	43,044	100	92
1981	54,195	9,619	63,814	100	85
1982	90,160	10,603	100,763	100	89
1983	125,886	14,461	140,347	100	90
1984	127,071	15,741[a]	142,812	100	89
1985	130,634	12,135[a]	142,769	100	92
1986	148,863	12,467[a]	161,330	100	92
Total	716,550	78,329	794,879	100	90

Sources: Nikkatsuren 1980–1984; Fisheries Agency of Japan 1984–1986.
[a]Suisan Shinchosha 1984–1986.

Single purse seine catches increased each year between 1980 and 1986 (Table 6). Catches rose from 39,741 tonnes in 1980 to 148,863 tonnes in 1986—an increase of 275 percent. The average vessel catch in the fishery, 3,312 tonnes in 1980, declined to 2,463 tonnes in 1981, then improved each year, reaching 4,802 tonnes in 1986. These data show that, along with aggregate increases in effort in the fishery, the rate of productivity per vessel continued to rise.

Catches by Japan's group purse seine fleet showed a similar upward trend. Total catches of the group purse seine fleet rose from 3,303 tonnes in 1980 to 15,741 tonnes in 1984, then declined to 12,467 tonnes in 1986 (Table 6). Average catches per group increased from 826 tonnes in 1980 to 2,249 tonnes in 1984, then eased to 2,078 tonnes in 1986.

The species composition of Japanese purse seine fleet catches is shown in Table 7 for the single purse seine fleet from 1980 to 1984 and for the group seine fleet from 1980 to 1986. For both fleets, skipjack tuna predominates. Between 1980 and 1984 the proportion of skipjack in single purse seine catches averaged 72 percent, ranging from 63 percent in 1981 to 80 percent in 1983. For the group purse seine fleet the percentage of skipjack catches ran even higher, averaging 81 percent between 1980 and 1986, with a range from 64 percent in 1981 to 97 percent in 1984.

Yellowfin tuna catches for the single purse seine fleet averaged 27 percent of total catches from 1980 to 1984, with other species averaging

Table 7. Species composition of Japanese single and group purse seine catches in the Pacific islands region 1980–1986

Single purse seine	Percentage[a]						
	1980	1981	1982	1983	1984	1985	1986
Skipjack	72	63	69	80	76	—	—
Yellowfin	25	36	29	19	24	—	—
Other	3	2	2	1	0	—	—
Total	100	100	100	100	100	100	100

Group purse seine	Percentage						
	1980	1981	1982	1983	1984[b]	1985[b]	1986[b]
Skipjack	74	64	78	85	97	74	93
Yellowfin	26	36	22	15	3	26	7
Total	100	100	100	100	100	100	100

Sources: Nikkatsuren 1980–1984; Fisheries Agency of Japan 1985–1986.
[a]Percentages may not add to 100 due to rounding.
[b]Suisan Shinchosha 1984–1986.

about 2 percent (Table 7). The proportion of yellowfin tuna catches for the group purse seine fleet between 1980 and 1986 averaged 19 percent. Group seiners do not report other species of fish caught because they are insignificant—1 percent—and have no commercial value.

U.S. Fleet

Data on catches by the U.S. purse seine fleet in the Pacific islands region are not available. However, because most tuna landed by U.S. vessels goes to U.S. canneries for processing, data on U.S. cannery receipts from the U.S. fleet in the western Pacific are used to approximate catches (Table 8). Table 8 does not include catches from vessels of foreign registry controlled by U.S. interests (for example, the Zee Enterprises fleet based in Guam).

Table 8 shows that between 1980 and 1985, the U.S. fleet in the Pacific islands harvested 501,322 tonnes of tuna. Annual catches increased from 11,435 tonnes in 1980 to a peak of 155,391 tonnes in 1984, then declined to 106,958 tonnes in 1985—an overall increase of more than 800 percent. Between 1980 and 1985, 70 percent of the catch (350,860 tonnes) was skipjack, 29 percent (145,833 tonnes) yellowfin, and 1 percent (4,629 tonnes) albacore and other species.[31]

Table 8. U.S. cannery receipts from U.S. fleet catches in the western Pacific Ocean by species 1980–1985

| | Tonnes | | | | | | |
	1980	1981	1982	1983	1984	1985	Total
Skipjack	10,129	16,999	35,158	94,960	113,772	79,842	350,860
Yellowfin[a]	985	12,010	20,072	45,203	41,134	26,429	145,833
Albacore	321	741	1,542	853	485	687	4,629
Total	11,435	29,750	56,772	141,016	155,391	106,958	501,322
Western Pacific catches as % of total fleet catches	6	15	30	60	66	53	

Source: USITC 1986, 161.
[a]Includes bigeye, blackfin, and bluefin tuna.

The data in Table 8 demonstrate the importance of the tuna fishery in the Pacific islands region to the U.S. fleet. In 1980, catches by the fleet in the western Pacific accounted for 6 percent of the total catch; by 1984 the proportion had risen to 66 percent. On average between 1980 and 1985, the U.S. fleet took 38 percent of its total catches in the Pacific islands; between 1983 and 1985 the proportion was much higher at 60 percent.

In terms of species harvested, 11 percent of the U.S. fleet's total skipjack catch was taken in the western Pacific in 1980; by 1985 the proportion had risen to 94 percent (Table 9). The proportion of yellowfin taken in the region was only 1 percent in 1980. By 1983 and 1984 the proportion had increased to 45 percent; it then fell to 24 percent in 1985. The relative importance of U.S. purse seine catches of albacore and other species in the western Pacific between 1980 and 1985 fluctuated widely, showing no clear trend.

The dramatic increase in catches by the U.S. purse seine fleet in the Pacific islands since 1980 is unparalleled in any other tuna fishery. Total catches doubled each year between 1980 and 1983 and have remained at a high level since then (Table 8).

Reductions in the size of the U.S. fleet—due to vessels being transferred to foreign flags or to other fisheries (for example, trawl fisheries) or being tied up owing to bankruptcy—and consequent declines in U.S. fishing effort in the region in 1985 and 1986 indicate that the record 1984 catch will not soon be surpassed. Nonetheless, the region's fishery will remain an important area of operation for U.S. vessels.

Table 9.   U.S. cannery receipts from U.S. fleet catches in the western
Pacific Ocean by species as a proportion of total U.S. cannery
receipts 1980–1985

	Percentage					
	1980	1981	1982	1983	1984	1985
Skipjack	11	21	42	74	86	94
Yellowfin[a]	1	11	20	45	45	24
Albacore	5	6	27	10	4	12

Source: USITC 1986, 161.
[a]Includes bigeye, blackfin, and bluefin tuna.

Other Fleets

Information about catches by other purse seine fleets in the islands
region is not readily available. However, from the number of vessels
operating and average catch rates, it can be estimated that 80,000 to
100,000 tonnes of tuna is harvested annually by other fleets. With the
growth in Korean and Taiwanese fleets and the transfer of more U.S.
vessels to foreign flags, the volume of tuna taken by other fleets will
increase.

OUTLOOK

The purse seine fishery in the Pacific islands region has become a major
international tuna fishery. Its growth since the mid-1970s has resulted
from both internal and external forces. Within the region, JAMARC's
research program, the fishery's productivity, and the willingness of
island governments to license DWFNs have fostered development. Outside the region, difficulties for the U.S. fleet in the eastern tropical Pacific
and the resulting deployment of vessels to the islands region led to
the fishery's rapid expansion.

The level of purse seine fishing effort by DWFNs in the Pacific islands region is expected to remain at a high level in the medium term,
though it is possible that island governments will join forces to regulate the future level of effort in the fishery so as to obtain a larger share
of the revenue generated by the fishery. Moreover, access for DWFN
purse seine fleets may be made conditional upon the transshipment
of tuna at ports in the region. The Law of the Sea Convention gives
coastal states the right to set such conditions, but island countries have
so far not exercised it.

If restrictions are imposed on DWFN purse seine fleets, Pacific island countries themselves are likely to invest in purse seine vessels or to implement policies designed to attract foreign vessels to base at their ports. For example, preferential fisheries access throughout the region may be given to vessels flying a Nauru Group country flag or to foreign vessels based permanently within the region. This possibility will be an important consideration for vessel owners if fishing effort is restricted and a queuing system for licenses is devised.

Setting limits on purse seine fishing effort may be expected to increase the profitability of all licensed purse seine vessels, especially vessels based in the region. One Japanese industry commentator has noted that if Papua New Guinea were to reduce the level of purse seine fishing effort within its EEZ, the price of skipjack tuna on the international market would rise. If a group of island countries (for example, the Nauru Group) were to agree to set limitations, the impact would be even greater. Initial losses in license revenue would be partly offset as access fees rose with the rise in fish prices. The effect of such a reduction in DWFN effort on the development of domestic fishing industries could be even more dramatic.

It can be argued that the effectiveness of measures to restrict fishing effort in the region will be eroded by the redeployment of vessels to other fisheries. Such an outcome is unlikely, however, because few purse seine fisheries are as accessible, as extensive, and as productive as the one in the islands region. Efforts to develop and expand the purse seine fishery in the Indian Ocean are likely to be sustained, but for most DWFN fleets it is locationally inferior to the Pacific islands fishery.

The desire of island governments to impose restrictions on purse seine fishing is likely to be reinforced by further attrition of the U.S. fleet. According to industry reports, the active U.S. purse seine fleet was 67 vessels in May 1986 (Felando 1986). There have been further reductions since then, and more are anticipated. Indeed, the U.S. purse seine fleet could stabilize at 50 to 55 vessels. If it does, the Pacific islands tuna fishery will become less important to the U.S. fleet because reductions in the size of the fleet could ease pressure on the resource in the eastern Pacific.[32] This reduction should lead to improved fishing conditions there, and possibly to a longer fishing season each year. Nonetheless, the tuna fishery in the islands region will remain a secondary fishery for the U.S. fleet because it is unlikely—for political and economic reasons—that the U.S. fleet will reestablish access to the EEZs of Latin American countries.

The purse seine fishery has been the Pacific islands' most controversial fishery. It will continue to be surrounded by controversy because the purse seine fishery is likely to be the first of the region's distant-water tuna fisheries to be subjected to licensing restrictions. Such restrictions will be imposed cooperatively as island countries seek to obtain a larger share of the benefits flowing from the exploitation of their tuna resources.

NOTES

1. I am indebted to Osamu Narasaki, president, New Fisheries Development, Ltd., Japan, who provided important background information for this paper.

2. In 1984 the size of the world tuna market was 78 million standard cases. Of this amount, 52 percent was consumed in the United States (King 1986, 58).

3. For example, the U.S. pole-and-line tuna fleet declined by 80 percent between 1970 (45 boats) and 1985 (9 boats) and the Japanese distant-water pole-and-line tuna fleet by 76 percent between 1973 (500 boats) and 1981 (120 boats). (USITC 1986, 157; Doulman and Wright 1983, 55).

4. Since 1963, Van Camp has been a wholly owned subsidiary of Ralston Purina, a diversified U.S. international corporation. Van Camp is the second largest U.S. tuna corporation, after Star-Kist.

5. Japanese fishermen had traditionally fished for tuna. Because in addition to the domestic Japanese market there was a resilient international demand for canning-grade tuna, the Japanese government gave high priority to the reestablishment and expansion of the country's tuna-fishing capability.

6. The Japanese already had a good understanding of the tuna resources in Micronesian waters. Before World War II, when Micronesia was a League of Nations-mandated territory of Japan, more than 7,600 Japanese fishermen had lived and fished commercially there (Matsuda and Ouchi 1984, 159). Between 1922 and 1939, tuna fishermen harvested about 124,000 tonnes of tuna (Purcell 1976, 206). Annual catches increased from 16 tonnes in 1922, peaked at 35,000 tonnes in 1937, and declined to 19,000 tonnes in 1939. In addition to tuna, Japanese fishermen caught significant quantities of other fish and shellfish.

7. Purse seine fishing based on U.S. technology was known in Japan as *bei-kin*, a Japanese abbreviation for "American-style purse seining" (Watanabe 1983, 1).

8. JAMARC continued exploratory fishing between 1977 and 1981 in the Pacific islands region but not with the degree of success it had

had around Micronesia and Papua New Guinea. Its vessel also operated in Kiribati, Solomon Islands, Marshall Islands north of New Caledonia, and the Indian Ocean (Watanabe 1983, 4).

9. For example, the Japanese developed deeper nets, about 80 percent longer than those normally used in the eastern tropical Pacific (Wankowski 1980, 20).

10. Table 1 shows 33 Japanese single seiners listed on the Forum Fisheries Agency's register of fishing vessels. Thirty-one of these vessels fish commercially, and two are research vessels operated by JAMARC.

11. For example, *Apollo*, *Zapata Pathfinder*, *Mary Elizabeth*, *Jeannette C*, *Voyager*, *Eastern Pacific*, and *Western Pacific*, among other U.S. vessels, undertook exploratory fishing in Papua New Guinea's EEZ between 1976 and 1980. Many of these seiners were associated with Star-Kist, which operated in Papua New Guinea's domestic tuna fishery.

12. In the early 1980s, U.S. vessels operating in the eastern Pacific were harvesting about 3,000 tonnes of tuna per year. After their migration to the western Pacific and adaptation to fishing conditions there, the vessels were averaging annual catches of 5,000 to 6,000 tonnes. Some vessels caught as much as 7,000 tonnes. Knowledge of these catch rates was a strong incentive for U.S. fishermen to relocate their operations.

13. JAMARC's research vessel, the *Nippon Maru* (999 GRT), was the one vessel on the register larger than 500 GRT.

14. The carrying capacity of the new 750-GRT seiners is about 750 tonnes each—a significant increase. Capacities of the 499-GRT seiners (depending on design) ranged from 450 to 550 tonnes.

15. Superseiners were designed—and many were constructed—when fuel prices were relatively low in the early 1970s and before extended jurisdiction was popularly introduced (1977 to 1980) around the world. However, fuel price increases and the impact of extended jurisdiction have eroded the operational advantages of superseiner-class vessels.

16. Two of the seiners are laid up in Pusan (South Korea); it is possible that they will be scrapped.

17. The South Korean government has been negotiating with Papua New Guinea for an access agreement since 1982. However, negotiations became deadlocked over wording in the agreement relating to Papua New Guinea's rights to detain Korean fishermen for

infringements of terms and conditions of access. The matter remains unresolved. However, because of Papua New Guinea's willingness to license DWFN fleets under the Japanese agreement, there was no urgency for Korea to conclude an agreement because its vessels would not be denied access.

18. According to industry reports, Taiwan's purse seine fleet is built by one shipyard (Fong Kuo Shipbuilding Co.). Its sister company (Fong Kuo Fishery Co.) currently operates one group seiner and three single seiners. The shipyard has the capacity to construct eight or nine seiners per year, with each vessel taking only four and a half months to complete. The average total cost of each vessel (including all equipment and gear) is $5.1 million (NT$180 million). Seiners built in Taiwan are identical to the Japanese seiners and about 15 percent cheaper. The Taiwanese vessels are equipped with Japanese engines, nets, and electronic fish-finding devices and U.S. (Marco) power blocks.

19. A subsidiary consideration is that JAMARC's exploratory purse seine fishing around Kiribati and the Marshall Islands was found to be less robust than fishing in the EEZs of the Federated States of Micronesia and Papua New Guinea (Watanabe 1983, 4–5).

20. For a brief description of group seine tuna fishing, see Doulman and Wright 1983, 54.

21. The group operates under special arrangements with Solomon Islands' government. In recent years it has made exceptional catches exceeding 6,000 tonnes per year.

22. U.S. purse seine skippers use code in their communications to try to disguise their locations, catch rates, and so on.

23. For a description of how this system operates in Papua New Guinea, see Doulman 1986c.

24. Although most agreements are with industry, DWFN governments play a prominent role in negotiating agreements. For example, negotiations with Japan are conducted by government officials, and industry representatives are rarely permitted to speak. In this respect, access agreements with Japan are essentially government-to-government agreements, even though the fishing associations, not the government, sign them.

25. The ATA had access agreements in the region from 1980 to 1984. The Federated States of Micronesia, the Marshall Islands, and Palau had an agreement from 1980 to 1982; Papua New Guinea in 1982; the Federated States of Micronesia, Kiribati, and Palau from 1982

to 1984; and the Cook Islands, Niue, Tokelau, Tuvalu, and Western Samoa from 1983 to 1984 (Doulman 1986b, 26). A treaty between the U.S. and 16 island governments providing for fisheries access to the region for the U.S. fleet is expected to take force in mid-1987.

26. As a percentage of total expenses before depreciation, access fee payments for U.S. purse seiners increased from 0.5 percent in 1980 to 0.9 percent in 1985 (USITC 1986, 164). Contrary to ATA arguments, access fee payments are an insignificant share of the cost of fishing operations.

27. For a more detailed discussion of this fleet, see Doulman 1986a, 6–8.

28. The United States is the world's largest canned-tuna processor. In 1985 it had eight canneries: one on the U.S. mainland (Terminal Island, California), five in Puerto Rico, and two in American Samoa (USITC 1986, 172). These canneries employed approximately 14,200 workers in 1985 at a total cost (wages and fringe benefits) of $120 million (USITC 1986, 173).

29. Most of the benefits from tuna transshipment at Tinian and Guam accrue to Guam because of its ability to service fleets. In most cases when fish is transshipped in Tinian, the seiners return to Guam to take on supplies, have their nets mended, and give crews a period of rest and recreation.

   Tuna is transshipped in Tinian primarily to circumvent U.S. domestic legislation (the Nicholson Act) relating to use of U.S. ports by foreign transport vessels. Foreign vessels can use Tinian but not Guam because of their different political statuses with the United States. Foreign transport vessels—especially those from Asian countries—frequently offer more competitive freight rates than their U.S. counterparts.

   Tuna transshipment has been a source of irritation for the governments of Guam and the Northern Marianas because, in the words of one U.S. official, "Guam gets all the benefits and Tinian all the pollution."

30. Based on information supplied by Star-Kist to the USITC, transshipment rates (including loading and unloading costs) in August 1986 were approximately $130 per tonne from Guam and Tinian to Puerto Rico and about $94 to American Samoa (USITC 1986, 190).

31. There is a significant difference between the species compositions of Japanese and U.S. single purse seine fleets' catches. Because U.S. seiners are larger, they can set deeper nets and harvest a higher proportion of nonsurface species of tuna. Hence U.S. seiners catch more—and larger—yellowfin than Japanese seiners do. In

standardized terms, the U.S. catch is more valuable than the Japanese catch. (Tuna for canning is graded by species, size, and other features such as the degree of salt penetration. Yellowfin are more valuable than skipjack, and larger yellowfin bring a higher price).

32. The overall outcome, however, will depend on actions taken by Latin American countries to develop their own purse seine fleets.

REFERENCES

Clark, Les.

1986   Tuna industry developments in the southwest Pacific. In Proceedings of the INFOFISH Tuna Trade Conference. Bangkok, Thailand. pp. 151–159.

Doulman, David J.

1986a   Options for U.S. fisheries investment in the Pacific islands. Pacific Islands Development Program. East-West Center. Honolulu. 26 pp.

1986b   Fishing for tuna: The operation of distant-water fleets in the Pacific islands region. Research report series no. 3. Pacific Islands Development Program. East-West Center. Honolulu. 38 pp.

1986c   Licensing distant-water tuna fleets in Papua New Guinea. Marine Policy 11(1):16–28.

Doulman, David J., and Andrew Wright

1983   Recent developments in Papua New Guinea's tuna fishery. Marine Fisheries Review 45(10):47–59.

Doulman, David J., and Robert E. Kearney

1986   The domestic tuna industry in the Pacific islands region. Research report series no. 7. Pacific Islands Development Program. East-West Center. In press.

Felando, August

1986   A perspective from the U.S. tuna seiner fleet: Has there been a change in policies by the U.S. government? Mimeo. 19 pp.

Fisheries Agency of Japan

1985–   Offshore section statistics. Tokyo.
1986

Franklin, Peter G.

1982   Western Pacific skipjack and tuna purse seine fishery: Development, current status, and future. Forum Fisheries Agency. Honiara, Solomon Islands. Mimeo. 30 pp.

Green, R. E., W. F. Perrin, and B. P. Petrich

1971   The American tuna purse seine fishery. In Modern fishing year of the world, vol. 3. Edited by H. Kristjonsson. Fishing News Books. London. pp. 182–194.

Habib, George

1984   Overview of purse seining in the South Pacific. Forum Fisheries Agency. Honiara, Solomon Islands. 26 pp.

King, Dennis M.

1986   The U.S. tuna market: A Pacific island perspective. Pacific Islands Development Program. East-West Center. Honolulu. 74 pp.

Matsuda, Y., and K. Ouchi

1984   Legal, political, and economic constraints on Japanese strategies for distant-water tuna and skipjack fisheries in Southeast Asian seas and the western central Pacific. In Memoirs of the Kagoshima University Research Center for the South Pacific 5(2):151–232. Kagoshima.

Nikkatsuren

1980–   *Katsuwo-maguro* yearbook (Skipjack and tuna yearbook).
1984   Tokyo.

Papua New Guinea

1982   Tuna. Fisheries Division. Department of Primary Industry. Port Moresby. Mimeo. 45 pp.

Purcell, David C.

1976   The economics of exploitation: The Japanese in the Mariana, Caroline, and Marshall Islands 1915–1940. Journal of Pacific History 11(3):189–211.

South Pacific Bureau for Economic Cooperation (SPEC)

1982   Summary record of the 13th South Pacific Forum. Report no. SPEC (82) REP. Suva, Fiji. 42 pp.

Suisan Shinchosha

1984–   *Katsuwo-maguro tsushin* (Skipjack and tuna report). Tokyo.
1986

United States District Court (Southern District of California)

1985   Ed Gann et al. versus Star-Kist Foods, Inc., Ralston Pu-
            rina, Inc., and Castle and Cooke, Inc. Civil action no.
            850553K(1). Los Angeles. 19 pp.

United States International Trade Commission (USITC)

1986   Competitive conditions in the U.S. tuna industry: Report
            to the president on investigation no. 332–224 under sec-
            tion 332 of the Tariff Act of 1930 as amended. Publication
            no. 1912. Washington, D.C. 320 pp.

Wankowski, J. W. J.

1980   Recent history and summarized statistics of the industrial
            fisheries for tuna and tunalike species in the area of the
            Papua New Guinea fisheries zone 1970-79. Research bulle-
            tin no. 24. Fisheries Division. Department of Primary In-
            dustry. Port Moresby, Papua New Guinea. 82 pp.

Watanabe, Yoh

1983   The development of the southern water skipjack tuna fish-
            ing grounds by the distant-water purse seine fishery.
            Bulletin of the Japanese Society of Fisheries Oceanogra-
            phy 42:36–40.

9.
Distant-Water Tuna Longline Fishery

Michael J. Riepen

INTRODUCTION

The distant-water tuna longline fishery in the Pacific islands region is one of the world's major fisheries. The combined landed catch value of the fishery is estimated to exceed $300 million per annum, with annual catch volumes exceeding 100,000 tonnes.

The waters of the Pacific islands host large populations of tunas: bigeye (*Thunnus obesus*), yellowfin (*Thunnus albacares*), skipjack (*Katsuwonus pelamis*), albacore (*Thunnus alalunga*), and southern bluefin tuna (*Thunnus maccoyii*). Some of these species form the basis of artisanal fisheries for Pacific islanders who for centuries have relied on tuna as a major food source. Since the end of World War II, the tuna populations in the Pacific islands have become the target of an expanding and very valuable commercial fishery for the industrialized distant-water fishing nations (DWFNs).

Japan entered the tuna longline fishery in the late 1940s. Today Pacific island countries have tuna longline fishing access agreements with Japan, Korea, Taiwan, and the Soviet Union. Some Pacific island countries have also developed their own longline industries. The longline fishery is the most important fishery financially in the islands region, contributing over two-thirds of the access fee payments made to the island countries.

This paper traces the development of the distant-water tuna longline fishery in the Pacific islands, highlighting trends for each of the major DWFNs. Economic and biological influences on the fishery are presented and likely developments discussed.

JAPAN

The development of a Japanese distant-water tuna longline fleet dates back to the middle and late 1940s. Before World War II, Japanese fishermen undertook longlining for tuna as a seasonal activity, using vessels designed principally for skipjack pole-and-line fishing. Fishing was

161

restricted to the coastal and offshore waters of the Japanese archipelago. Catches were chilled with ice because freezing equipment was not yet installed on vessels.

Within a couple of years after the end of World War II, Japanese distant-water fishermen started to invest in larger vessels. The industry expanded rapidly during the 1950s with the 1952 lifting of the MacArthur Line and the development and installation of freezing equipment aboard vessels.

The operational strategy of the Japanese distant-water tuna longline fleet during the 1950s and early 1960s was based on supplying fish to the international—notably the U.S.—tuna-canning market. Mothership operations were adopted in the early 1950s. The economic advantage of establishing shore bases was quickly realized, however, and bases were soon established in American Samoa (1954), Vanuatu (1957), and Fiji (1963). The target species of this fishery were albacore and yellowfin.

Fishing effort and catches by Japanese longliners working from Pacific island bases increased steadily, reaching a high of about 120 vessels. These vessels caught 30,000 to 40,000 tonnes of tuna annually between 1962 and 1965 (Tables 1 and 2). After 1965 the number of Japanese longliners fishing in the region fell dramatically; by 1970 almost

Table 1. Distant-water tuna longline vessels operating in the Pacific islands region for the international canning market 1955–1985

	Japan	Korea	Taiwan	Other	Total
1955	60	0	0	0	60
1960	70	5	0	1	76
1965	120	30	30	0	180
1970	10	100	130	0	240
1975	0	200	200	0	400
1980	0	180	150	0	330
1985	0	60	50	7	117

Sources: Wetherall and Yong 1984; Clark 1986.

all Japanese vessels had withdrawn from the fishery. This decline was caused by the rapid expansion of the Japanese economy, escalating operational costs (especially wages), a stagnant and weakening market for canned tuna, and the development of ultra–low-temperature freezing technology. Although Japan withdrew from longlining for the canning market, aggregate fishing effort and catches did not decline because Taiwanese and Korean vessels replaced Japanese vessels in the fishery (Table 3).

Table 2. Estimated annual catches in tonnes of the distant-water tuna longline fleets in the Pacific island region for the international tuna canning market 1955–1985

	Tonnes				
	Japan	Korea	Taiwan	Other	Total
1955	10,000	0	0	0	10,000
1960	28,000	2,000	0	0	30,000
1965	25,000	3,000	8,000	0	36,000
1970	6,000	14,000	19,000	0	39,000
1975	0	21,000	26,000	0	47,000
1980	0	16,000	23,000	0	39,000
1985	0	14,000	12,000	2,000	28,000

Sources: Wetherall and Yong 1984; Clark 1986.

Table 3. Number of vessels listed on the Regional Register of Fishing Vessels by fishing method and by country in 1986

Country	Fishing method			Total
	Longline	Pole-and-line	Purse seine	
Japan	994	122	72	1,188
Korea	122	0	14	136
Soviet Union	9	0	13	22
Taiwan	171	0	12	183
United States	0	0	62	62
Solomon Islands	2	0	0	2
Others	29	8	33	70
Total	1,327	130	206	1,663

Source: Terawasi 1986.

The development of ultra–low-temperature (below –50°C) freezing machinery paralleled the decline in the Japanese tuna longline canning fleet. Fast freezing was essential for producing high-quality tuna for Japan's sashimi market, until then supplied by local Japanese vessels using ice. Japanese vessel operators quickly installed the freezing equipment when it became available in the 1960s, and by 1970 a new Japanese longline fleet had developed and was fishing throughout the Pacific islands from Palau to Kiribati and southward to Australia and New Zealand. This fleet, which targeted on southern bluefin tuna, bigeye, yellowfin, and billfish, was based in Japan. Its catch was carefully handled, deep-frozen, and sold exclusively on the sashimi market.

The Japanese sashimi tuna longline fleet is now the single largest foreign fleet fishing in the Pacific islands (Table 3). More than half the fishing access fee payments made to Pacific island countries have historically come from this fleet.

During the 1970s the number of vessels in the Japanese sashimi tuna longline fleet remained reasonably stable at 1,400 to 1,600 vessels operating around the world. The fleet's annual catches fluctuated around 300,000 tonnes, 25 to 35 percent taken in the Pacific islands region.

In the middle and late 1970s, Korean and Taiwanese distant-water longline operators, earning poor returns from harvesting for the canning market, entered the sashimi longline fishery. Their entry was often facilitated by the use of secondhand Japanese vessels or financial support from Japanese trading companies or both. But this expansion in the fishery, coupled with the development of ultra-cold reefer vessels in the early 1980s, rapidly oversupplied the Japanese sashimi market and depressed prices, especially for the lower-grade *akami* tuna. This situation, linked to an already stagnant and contracting sashimi market, induced Japanese authorities to develop strategies for improving the economic viability of the country's distant-water tuna longline fleet. Research programs were initiated to improve vessel design and fishing techniques, market promotion was expanded, and the size of the fleet was reduced. A 20 percent cut in fleet capacity (measured by gross registered tonnage) was implemented between 1981 and 1983 by withdrawing vessels from the fishery and not replacing them. Currently there are about 1,200 operational Japanese sashimi longline vessels, the majority of them in the range of 20 to 80 gross registered tonnes (GRT) or over 200 GRT (Table 4). Most of these vessels are listed on the Forum Fisheries Agency's Regional Register of Fishing Vessels.

Catches by the Japanese sashimi tuna longline fleet in the Pacific islands total an estimated 90,000 tonnes annually. These catches consist of yellowfin (40,000 tonnes), bigeye (30,000 tonnes), southern bluefin (10,000 tonnes), and billfish and other species (10,000 tonnes). The distribution of catch for the two major species, yellowfin and bigeye, for the years 1982 to 1985 (Figures 1 and 2) illustrates the importance of the equatorial oceanographic circulation system. The distribution of catch is relatively uniform in areas outside this system.

Catch rates (measured by the number of fish caught per 100 hooks) for the fishery have declined over time. Southern bluefin tuna and

Table 4. Number of longline vessels listed on the Regional Register of Fishing Vessels by size class and by country in 1986

	Size class (GRT)				
Country	60	60–100	100–200	200	Total
Japan	332	193	104	365	994
Korea	3	0	9	110	122
Soviet Union	0	0	0	9	9
Taiwan	79	1	72	19	171
Solomon Islands	0	0	2	0	2
Others	0	0	0	29	29
Total	414	194	187	532	1,327

Source: Terawasi 1986.

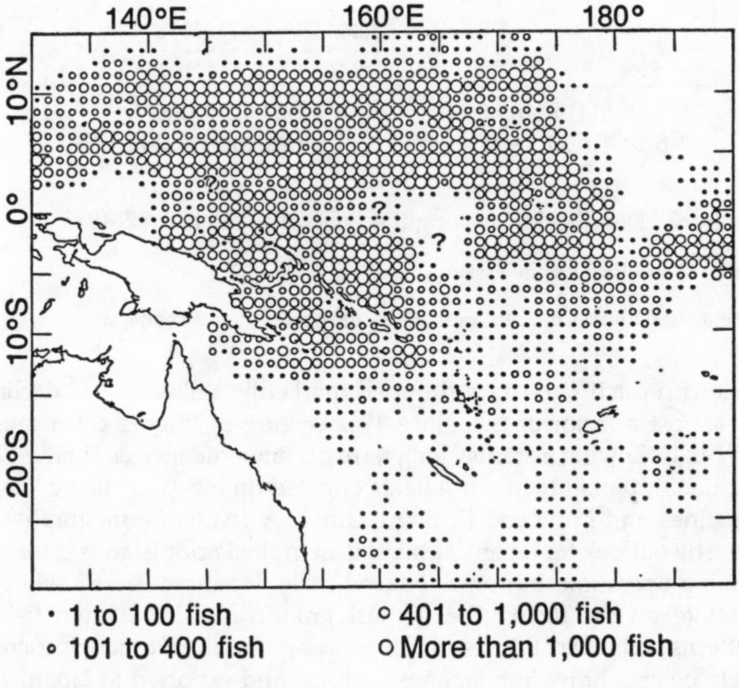
140°E
160°E
180°
10°N
0°
10°S
20°S
?
?
?
• 1 to 100 fish
∘ 101 to 400 fish
○ 401 to 1,000 fish
○ More than 1,000 fish

Figure 1. Distribution of longline bigeye catch 1982–1985[a]

Source: South Pacific Commission.

[a]The question marks indicate areas where data coverage is incomplete.

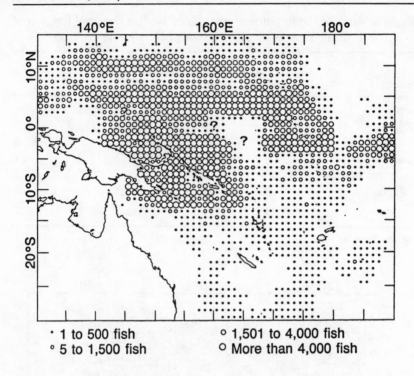

* 1 to 500 fish   o 1,501 to 4,000 fish
o 5 to 1,500 fish   O More than 4,000 fish

Figure 2. Distribution of longline yellowfin catch 1982–1985[a]

Source: South Pacific Commission.

[a]The question marks indicate areas where data coverage is incomplete.

yellowfin catch rates have dropped markedly. Yellowfin has declined by almost a factor of two since 1962 (Figure 3). Bigeye catch rates— although showing a gradual long-term decline—did increase in the 1970s (Figure 3), probably because fleets targeted on bigeye by using "deep" longlines and improved fishing technology (bathythermographs).

The outlook for Japan's sashimi fleet in the Pacific islands is for continuing operation, but on a reduced scale. Japanese vessels will concentrate on producing the highest grade of sashimi tuna; fishing patterns will reflect this trend. The lower-grade *akami* tuna will increasingly be caught by non-Japanese vessels and exported to Japan. Furthermore, the cost of operating distant-water fishing vessels will continue to work to Japan's disadvantage. Japan will try to streamline vessel operations by improving fuel efficiency and propulsion systems, increasing mechanization to reduce manpower needs, and raising productivity by the use of new fishing technologies.

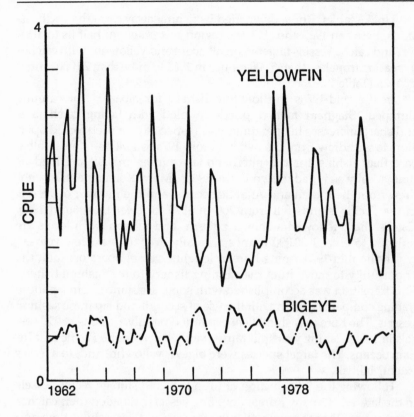

Figure 3. Trends in catch per unit of effort (CPUE) for bigeye and yellowfin by longline vessels in the area from 10°N to 15°S and 140°E to 180°W 1982–1985[a]

Source: South Pacific Commission.

[a]CPUE is measured in fish caught per 100 hooks.

KOREA

Korea was the second DWFN to enter the tuna longline fishery in the Pacific islands region. Korea began exploratory distant-water tuna long-lining in the Indian Ocean in 1957 and commercial fishing in the Pacific islands region soon thereafter. The expansion of the Korean fleet through the 1960s and early 1970s was dramatic. The number of vessels grew from none in 1955 to 30 in 1965 to a maximum of 200 in 1975 (Table 1). Almost all these vessels fished for the international canning market, supplying the bases operated by Japanese and U.S. trading and canning companies, including those in American Samoa, Fiji, and Vanuatu.

Unlike Japan, Korea diversified its fishing away from the Pacific islands; between 1965 and 1975 the region accounted for half its total effort and catch. Vessels targeted on albacore and yellowfin, with catches increasing from less than 5,000 tonnes in 1965 to more than 20,000 tonnes in 1975 (Table 2).

In the mid-1970s the longline fishery for albacore for canning slumped. Stagnant market prices coupled with falling catch rates and sharp increases in operating costs (especially for fuel) forced operators to withdraw from the fishery. From 1975 to 1986 the Korean albacore fleet fishing for the international canning market declined as dramatically as it had grown in the 1960s. Many Korean vessels withdrew from the Indian and Atlantic oceans, and the number of vessels in the Pacific islands fell from 200 in 1975 to fewer than 60 in 1986. Catches have followed a similar pattern, and Korean catch levels are unlikely to exceed 20,000 tonnes annually over the next few years.[1]

In the mid-1970s some Korean longline vessel operators took the opportunity to move from the canning fishery to the sashimi fishery. This transition was accomplished with some assistance from Japanese trading companies, often with the use of secondhand Japanese sashimi vessels. The Korean sashimi fleet expanded rapidly, peaking at 219 vessels in 1980. Fishing was split fairly evenly between the Pacific and Indian oceans. The target species were bigeye, yellowfin, and, to a lesser extent, billfish.

Following the introduction of ultra-low cold storage reefer vessels in the late 1970s, many Korean longline vessels offloaded their catches directly onto those vessels, thereby increasing their fishing time and reducing nonproductive operational costs. This practice contrasted with that of the Japanese fleet, which unloaded its catches directly onto the market in Japan at the end of each trip. It was not until the early 1980s that Japanese sashimi fishing vessels used the ultra-cold storage reefer services regularly.

Since 1980, because of restrictions on the export of secondhand Japanese sashimi vessels and the stagnant and weakening market for *akami* tuna, the Korean fleet has steadily declined. By 1985 the fleet was down to 156 vessels, 58 operating in the Pacific islands, 60 in the Indian Ocean, and 38 in the Atlantic Ocean.

Catches by the Korean sashimi tuna longline fleet in the Pacific islands were estimated at about 12,000 to 15,000 tonnes in 1980, falling to 7,000 to 10,000 tonnes in 1985.

The outlook for the Korean fleet is for possible stabilization in the short term and modernization in the medium term. Because of its cost competitiveness relative to Japan's, the Korean fleet is expected to maintain, even possibly to increase, its share of the Japanese sashimi market as newer, more modern vessels are introduced to the fleet. The number of vessels operating in the Pacific islands region will depend largely on catch rates and how they compare with rates in fishing grounds in the Atlantic and Indian oceans. At a minimum, the fleet is likely to continue operations in the Pacific islands at a level close to current operations.

TAIWAN

Taiwan entered the distant-water tuna longline fishery in the early 1960s. Like Japan and Korea, it directed its initial efforts to supplying the international tuna-canning market. Taiwanese longline fishing in the Pacific islands commenced in 1963, supplying albacore and yellowfin to the American Samoa canneries. Fishing effort and catches in the region rose sharply, peaking in 1973 with a fleet of around 200 vessels and an annual catch of more than 25,000 tonnes (Tables 1 and 2). The Taiwanese fleet operated throughout the Pacific islands, landing its catch at bases in American Samoa, Fiji, and Vanuatu.

The sharp increase in Taiwan's longline fishing effort in the late 1960s and early 1970s resulted from the redeployment of vessels from other fisheries combined with the availability of secondhand tuna longline vessels and related financing from Japan. Because of the two oil price hikes in the early and mid-1970s and a weakening market for albacore, Taiwan's longline fishery declined steadily. Fishing effort in the Pacific islands fell significantly. Vessels either moved into the Atlantic and Indian ocean fisheries seeking improved catch rates or retired from the fishery. By 1985 the Taiwanese longline fleet supplying fish to the international tuna-canning market was around 100 vessels, half of them operating in the Pacific islands. Unlike Korea, Taiwan continued operating from the three shore bases in the region: American Samoa, Fiji, and Vanuatu. Korean vessels withdrew from Vanuatu in the mid-1970s and from Fiji in the early 1980s. Taiwanese longline vessels now operate only from American Samoa.

Taiwanese tuna longline vessels operating for the international tuna-canning market, like Korean vessels, tend to concentrate fishing effort in the more subtropical areas of the Pacific islands (Figure 4). However, when Taiwan operated from Fiji and Vanuatu, the fleet's fishing effort was more widespread, extending toward Australia.

The shore bases in Fiji and Vanuatu were important unloading ports for the Taiwanese longline fleet. Both bases were owned and operated by Japanese trading companies. But when vessel numbers fell in the 1980s, their economic viability became uncertain, and ownership was transferred to the Fiji and Vanuatu governments. These governments

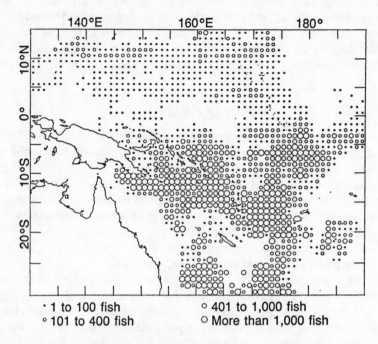

* 1 to 100 fish   ○ 401 to 1,000 fish
○ 101 to 400 fish   ○ More than 1,000 fish

Figure 4. Distribution of albacore longline catch 1982–1985

Source: South Pacific Commission.

will probably continue to operate the bases, seeking to involve commercial companies in their management and ownership if they can find suitable partners.

The total catch of the Taiwanese longline fleet for the international canning market in 1985 was about 10,000 tonnes; production will probably continue around that level. In the medium term, increased fishing effort by the Taiwanese fleet is likely to displace effort from competing DWFN fleets because of its lower operational costs, especially for crew. Taiwanese longliners will probably be employing more and more Pacific islanders.

Taiwan, like Korea, entered the Japanese sashimi market in the late 1970s. Taiwan's fishing effort was concentrated in the Indian Ocean, with only limited fishing in the Pacific islands region. This trend is expected to continue, with Taiwanese longline vessels supplying yellowfin and bigeye to the sashimi market through the use of ultra-low-temperature reefer services.

OTHER DISTANT-WATER FISHING OPERATIONS

The Soviet Union and some Pacific island countries also participate in the distant-water tuna longline fishery in the Pacific islands. The Soviet Union entered the fishery in 1985 and 1986, but its record of performance has not been good. In 1985 it had a fleet of five to ten tuna longliners operating in Kiribati's waters for yellowfin, bigeye, and albacore. Annual catches by these vessels were estimated to be under 2,000 tonnes. Kiribati did not renew Soviet fishing access in 1986; however, Vanuatu concluded a fishing access agreement with the Soviet Union in 1987. About as many Soviet vessels are expected to fish in Vanuatu as operated earlier in Kiribati. Fishing effort and catch by Soviet longline vessels in the Pacific islands will tend to be variable and significantly influenced by political considerations. However, the Soviet Union appears to be serious about expanding its fisheries. Consequently, increases are expected in fishing effort and catch levels by the Soviets in the Pacific islands.

Pacific island countries themselves have developed a distant-water longline fishing capability. Solomon Islands operates two sashimi tuna longline vessels and Tonga an albacore longline vessel. These vessels harvest tuna in the exclusive economic zones of neighboring countries. Their catches are low, not over 1,000 tonnes per annum. Although the development of longlining by Pacific island countries has not progressed rapidly over recent years, partly because of the depressed global tuna market, increased development can be expected. The recent acquisition of onshore storage and processing facilities in Fiji and Vanuatu will stimulate this development. It will require substantial input of capital and technical skills, which are most likely to come from the nations that have had a strong distant-water fishing presence in the Pacific islands in the past.

NOTE

1. Albacore catches in the Pacific islands are not uniformly distributed throughout the area; more albacore is caught in the subtropical areas of the region. See Figure 4. However, albacore catch rates have declined significantly since 1962 (Figure 5). Recent research indicates that similar yields could be taken from the fishery with a lower level of fishing effort (Wetherall and Yong 1984).

REFERENCES

Clark, Les

1986   South Pacific oceanic fisheries. Report 85/2. Forum Fisheries Agency. Honiara, Solomon Islands. 18 pp.

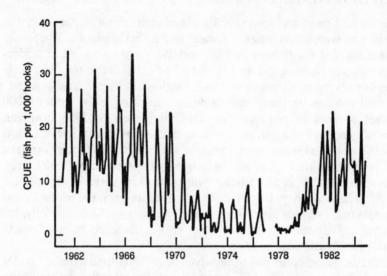

Figure 5. Trends in catch per unit of effort (CPUE) for albacore by longline vessels in the area from 5°N to 30°S and 145°E to 180°W 1962–1985

Source: South Pacific Commission.

Farman, R., and J. Sibert

1986   A review of southern albacore catch data from the South Pacific Commission region. Paper presented at the First South Pacific Commission Albacore Research Workshop. Auckland, New Zealand. 9 pp.

South Pacific Commission Tuna Program

1985   Yellowfin tuna catch rates in the western Pacific. Paper presented to the Seventeenth Regional Technical Meeting on Fisheries. South Pacific Commission. Noumea, New Caledonia. 33 pp.

1986   Tuna stocks of the southwest Pacific. Paper presented to the Eighteenth Regional Technical Meeting on Fisheries. South Pacific Commission. Noumea, New Caledonia. 33 pp.

Terawasi, Peter

1986   The Regional Register of Fishing Vessels. Report 86/55. Forum Fisheries Agency. Honiara, Solomon Islands. 10 pp.

Wetherall, J., and M. Yong

1984   Assessment of the South Pacific albacore stock based on changes in catch rates of Taiwanese longliners and estimates of total annual yield from 1964 through 1982. Report H–84–11. Southwest Fisheries Center. National Marine Fisheries Service. Honolulu. 7 pp.

    ... , ... , ...

PART III. ARTISANAL AND DOMESTIC TUNA FISHERIES

10.
The Importance of Small-Scale Tuna Fishing: A Tokelau Case Study

Robert Gillett and Foua Toloa

INTRODUCTION

Tuna is important to small-scale fishermen in the Pacific islands, but detailed information is hard to find. Clark (1986) says that "data on the subsistence fisheries and on the relative importance of tuna for subsistence fishing people is sparse." This scarcity of information is of concern to fishery scientists because it impedes analysis of the impact of industrial tuna fisheries on the operations of small-scale tuna fishermen (South Pacific Commission 1984, 1985).

Studies of indigenous tuna fisheries in Polynesia, Micronesia, and Melanesia have not been undertaken to gauge the magnitude of tuna catches relative to other fish catches nor to assess their nutritional or social value.[1] But the need for this type of research is recognized. The 1984 Food and Agriculture Organization's World Fisheries Conference— attended by delegations from nine Pacific island countries—urged that steps be taken to augment information about the contribution of subsistence fisheries to food supply (Food and Agriculture Organization 1984). A 1986 study of fisheries research needs in the Pacific islands recommended that information on traditional resource-use patterns be assembled as background for management of subsistence fisheries (Fakahau and Shephard 1986).

The Setting

A New Zealand dependency since 1924, Tokelau is made up of three low-lying atolls. These atolls—Fakaofo, Nukunonu, and Atafu—lie between 427 and 509 kilometers to the north of Apia, Western Samoa (Figure 1). Neighboring island groups include Tuvalu to the west, the Phoenix Islands (Kiribati) to the north, and the Cook Islands to the east. All the Tokelau atolls are closed; there are no passes through the reef separating the lagoon from the ocean. Small craft can cross the reef at about 15 natural depressions and two blasted channels, but the

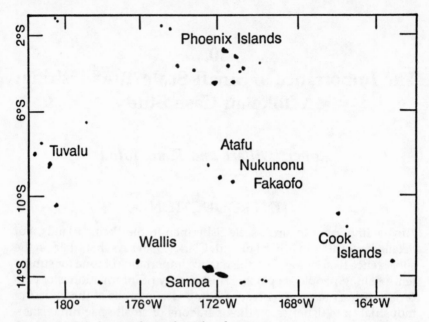

Figure 1. Tokelau and nearby island groups

trip is usually hazardous, even in a moderate swell or wind chop. There are no safe anchorages for ships, and only rarely do vessels attempt to anchor outside the reef.

Fakaofo, the site of this study, consists of approximately 60 small islets (Figure 2). About 4 square miles of land together with a 32-kilometer reef encircle 59 square kilometers of lagoon. The atoll is populated by about 700 Polynesian people, the vast majority living at Fale (the original village) or at Fenuafala (inhabited since 1958).

The Fakaofo people have always depended on the marine environment for food. With coconut and pandanus the only local food plants at the time of first European contact (Hooper 1984), Tokelauans have had to be good fishermen (Gillett 1985).

Tuna Fishing

Aspects of tuna fishing at Fakaofo have been recorded by Macgregor (1937), Kirifi (n.d.), Hooper (1984), and Gillett (1985). In the past 50 years, seven techniques have been used regularly to catch tuna. Currently, four of these methods produce almost all of the landings: handlining using rock and chum, shallow handlining using chum, deep handlining, and trolling using outboard engines. Figure 2 shows the primary tuna fishing areas around Fakaofo.

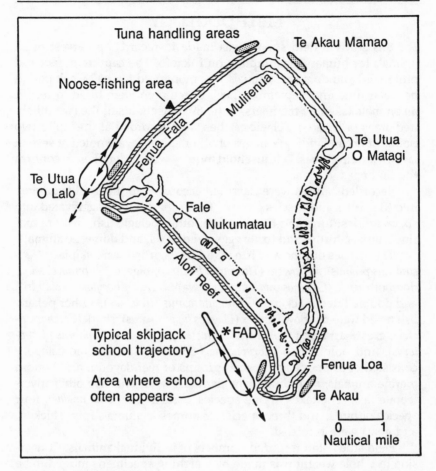

Figure 2. Fakaofo Atoll

Previous Catch Studies

Gulbrandsen (1977) estimated that an annual total of 28,181 kilograms of all species of fish was required to satisfy the nutritional demands of the 665 Fakaofo residents. Hooper (1984) monitored all fish catches on Fakaofo for a five-week period in 1981 and reported a weekly catch of about 1,545 kilograms.

Gillett (1985) recorded all pelagic fish catches over a five-week period in 1985 and determined that 435 individual tuna and tunalike fish and 431 other large pelagic fish were captured in the 114 fishing trips made by villagers.

METHODOLOGY

For this study a field survey was made to record the harvest of all animals for human consumption on Fakaofo. The capture of fish and other wild and domestic animals was monitored for a 12-week period between June and September 1986. Three men were hired to record all animals taken by members of the 69 households of the two inhabited areas of Fakaofo. Pelagic catches were recorded at the atoll's two points of ocean landing; catches of other species were noted at several lagoon landing areas. A household inquiry was undertaken to confirm the findings.

Recorded catches were classified according to the local taxonomy into 63 species and species groups. These species were aggregated into six categories: tuna and tunalike fish, other pelagic fish, inshore marine animals, turtles and turtle eggs, wild birds, and domestic animals.

The species of tuna and tunalike fish recorded were skipjack (*Katsuwonus pelamis*), yellowfin (*Thunnus albacares*), bigeye (*Thunnus obesus*), dogtooth tuna (*Gymnosarda unicolor*), wahoo (*Acanthocybium solandri*), and double-lined mackerel (*Grammatorcynus bilineatus*).[2] Other pelagic fish noted were rainbow runner (*Elagatis bipinnulatus*), shark (*Carcharhinidae*), great barracuda (*Sphyraena barracuda*), sailfish (*Istiophorus platypterus*), and dolphinfish (*Coryphaena hippurus*). The inshore category consisted of 39 species or species groups of reef, lagoon, and bottom marine animals, including giant clam (*Tridacna*) and five other invertebrate taxa. The takes of one species of turtle (*Chelonia mydas*), four species of birds, and three species of domestic animals (pig, chicken, and goat) were recorded.

Field study staff recorded numbers of individual animals. Conversion to whole weight was made by weighing specimens judged to be of average size. Catches of all pelagic species were increased by 2 percent to compensate for fish eaten during trips or used as bait. The reported catch of inshore species was increased by 10 percent to account for fish used at sea, for clandestine fishing activity (fishing on Sundays, for example), and for fishermen arriving during nonmonitored periods (that is, during the night). Because coverage for the taking of turtles, wild birds, and domestic animals was considered to be 100 percent, their values were not adjusted. However, the collection of turtle eggs—a discouraged practice—was estimated.

FIELD SURVEY RESULTS

Figure 3 and Table 1 show the results of the field survey.

In terms of whole-animal weight, tuna and tunalike species constituted 19 percent of Fakaofo animal food taken for human consumption

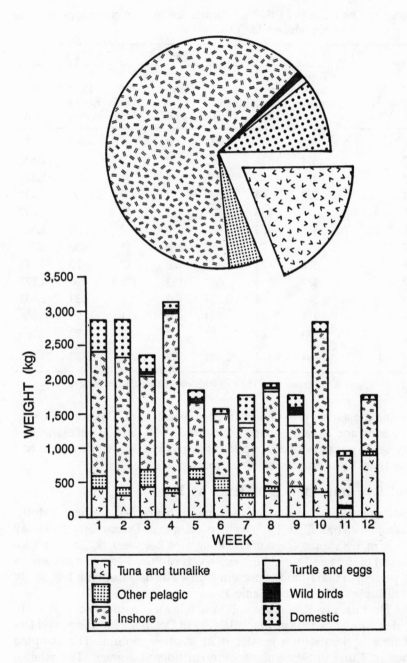

Figure 3. The taking of Fakaofo animals for human consumption
June 23–September 13, 1986

Table 1. The taking of Fakaofo animals for human consumption June 23 to September 13, 1986

Week	Tuna and tunalike fish	Other pelagic fish	Inshore fish	Turtles and eggs	Wild birds	Domestic animals	Total
				Category[a]			
1	422	177	1,814	0	1	460	2,874
2	314	116	1,892	0	1	557	2,880
3	436	255	1,355	0	51	261	2,358
4	350	64	2,570	0	20	130	3,134
5	549	148	971	0	41	131	1,840
6	377	183	936	0	2	70	1,568
7	273	73	959	28	11	430	1,774
8	370	73	1,383	28	17	70	1,941
9	437	6	881	153	105	190	1,772
10	349	6	2,334	0	4	139	2,832
11	114	37	734	0	1	70	956
12	893	49	752	0	5	70	1,769
Total	4,884	1,187	16,581	209	259	2,578	25,698

[a]Whole weight in kilograms. The shell weight was subtracted for *Tridacna*.

during the survey period, varying between 11 and 50 percent a week. The tuna and tunalike species consisted of 57 percent yellowfin and bigeye, 23 percent wahoo, 16 percent skipjack, 4 percent dogtooth, and less than 1 percent double-lined mackerel.

Relative Importance of Tuna as Food

To substantiate the accuracy of the tuna catch recorded during this study, Fakaofo data by Gillett (1985) were re-examined. During Gillett's 35-day study in 1985, an average of 505 kilograms per week (222 to 755 kilograms) of tuna and tunalike species was captured—an average within 10 percent of that from the present study. The data from the 1985 study thus support the data in Table 1.

The quantity of local animals available for consumption in Fakaofo varies throughout the year; availability in October, November, and December is significantly greater than in other months. The sampled months (June to September) were considered average. The relative proportions of the six food categories are thought to remain constant throughout the year.

To determine the actual dietary contribution of tuna, the study also considered food imports and exports. It is estimated that 400 kilograms of fish is exported overseas from Fakaofo on each voyage of the monthly ship. To determine the imports of animal foodstuffs, shipping records of the Office for Tokelau Affairs were examined. For 1986, Fakaofo animal foodstuff imports consisted of 941 kilograms in July and 1,047 kilograms in August of canned fish, canned and frozen meat, powdered and liquid milk, canned meat dripping, eggs, and butter. Tokelau officials considered these average amounts. In addition, the ship's passengers bring food into Fakaofo. Residents typically bring in 400 kilograms of frozen meat a month. Thus a crude estimate of animal food imports for the period June to September is 1,400 kilograms a month.

To gauge the relative importance of the various types of animal foods, the actual food yield was also considered. Tokelau-style preparation of several fish during the field study showed that about 85 percent of the whole-fish weight is customarily consumed directly.[3] The food yield for whole fish is probably greater than that of imported frozen meats but less than that of canned products. In this study, the food yield from local and from imported animal sources was assumed to be equal.

On the basis of reliable data for Fakaofo animal food production and less reliable information about imports, it appears that about 85 percent of the animal food consumed on Fakaofo was from local sources. Tuna and tunalike fish constituted about 16 percent of the total food intake.

Social Aspects

Hooper (1984) says that in Tokelau "fish are more than just food." On Fakaofo the point may taken further: tuna are more than just fish. Status, dietary preference, and recreation figure in the popularity of tuna.

Skilled tuna fishermen on Fakaofo enjoy special status in the community. They are believed to be smarter and stronger than fishermen who fish in the lagoon. Their status is enhanced because tuna fishing is a topic of everyday discussion and is highly visible, especially during the capture and unloading of fish. Furthermore, because skipjack catches are divided among the entire community, its members are interested in the productivity of tuna fishermen.

Taste preferences for cooked fish vary. For eating raw, however, most Tokelauans prefer tuna. Large chunks of boneless tuna are especially appreciated.

Tuna fishing in Tokelau also has recreational value. In fact, skipjack fishing was once "the greatest sport of Tokelau men" (Macgregor 1937). Because several men can handline for tuna from one boat and

several boats can fish in the same vicinity, the activity has a special social character. The men enjoy joking and lighthearted insults, especially
during the heated action of a big hookup and the fumbling that goes
with it. The intense competitiveness in tuna fishing is an important
part of Tokelaun social interactions. Finally, the thrill of catching, for
example, one noble 20-kilo tuna is thought to be superior to catching
40 less noble half-kilo creatures.

TUNA FISHING ELSEWHERE

The tremendous diversity of economic and geographic conditions in
the Pacific islands makes it difficult to extrapolate the results of the
Fakaofo study directly to other areas in this region. For example, because of Tokelau's association with New Zealand, the amount of money
flowing into Fakaofo is probably greater than it is for most small islands
in the region.[4] This association probably encourages food imports and
decreases the relative importance of tuna.

There is no quantitative data on small-scale tuna fishing in many
Pacific islands. Where information has been collected, it is often in non-
published form or difficult to locate. Table 2 summarizes data on various small-scale tuna fisheries throughout the Pacific islands region. The
information in the table is probably only a small portion of what would
be available to an enterprising researcher.

Comparisons between countries are complicated not only by varying degrees of reliability of existing information but also by nomencla-
ture: there is no standard of usage for such terms as *small-scale*,
subsistence, artisanal, indigenous, traditional, noncommercial, or even the
names of species in the various categories of tuna, ocean fish, or offshore pelagics.

In the Pacific, small-scale fishing is more important to residents of
islands with limited land resources than it is to people on larger, more
fertile islands. Tuna is more important than other fish where lagoon
resources are limited or absent. Taste preferences, geographic isolation,
and traditional values may also affect the level of tuna fishing.

A major shortcoming of the data in Table 2 is that they do not re-
veal the human aspect of what may seem to be meager quantities of
fish. Gillett (1987) tells how some Pacific islanders feel about tuna:

> There is a strong heritage of tuna fishing at Satawal (central Caro-
> line Islands). Although its soil is more fertile than that of most coral
> islands, the number of plants which can be cultivated is extremely
> limited. Taro and breadfruit make up most of the diet. Fish pro-
> vides a welcome change of food; however, due to the lack of a

Table 2. Tuna catch information from various small-scale fisheries in the Pacific islands

Country	Information
Cook Islands	A field survey determined in 1978 that skipjack and other ocean fishes contributed the following amounts to the artisanal fishery: Rarotonga 7 percent, Aitutaki 0 percent, Mangaia 13 percent, Atiu 40 percent, Mauke 51 percent (Cook Islands Ministry of Agriculture and Fisheries 1979).
	It is estimated that tuna in 1986 was 53 percent of the total small-scale fish catch of 300 tonnes (Neil Sims, personal communication 1986).
Fiji	The small-scale catch of tunas in 1986 was insignificant (Anthony D. Lewis, personal communication 1986).
Kiribati	On the basis of household interviews, the percentage of tuna in the predominantly subsistence fishery was determined for the following islands: Abemama 5 percent, Aranuka 21 percent, Arorae 30 percent, Kuria 43 percent, North Tarawa 3 percent, and Temana 62 percent (Mees n.d.).
	Skipjack and yellowfin made up 81 percent of the total artisanal fish catch of 941 tonnes on South Tarawa in 1977-78 (Crossland and Grandperrin 1979).
	Indigenous fishermen caught 12,304 kg of skipjack in the South Tarawa region in 1981 (Kiribati Ministry of Commerce and Industry n.d.).
Marshall Islands	Fish co-op data show that tuna were approximately 23 percent of the total of about 459 tonnes caught by small-scale fisheries on Majuro Atoll (Robert Carpenter, personal communication 1986).
New Caledonia	Skipjack and other tuna catches by the artisanal fleet in 1977 totaled 186 tonnes (Crossland and Grandperrin 1979).
Niue	Approximately 35 tonnes of tuna and tunalike species were caught in the 12-month period between November 1985 and October 1986 (Bradley Punu, personal communication 1986).

Table 2. *(continued)*

Country	Information
Solomon Islands	Tunas accounted for 12 percent of total artisanal fish catches in 1977 (Crossland and Grandperrin 1979).
	Skipjack and yellowfin accounted for approximately 10 percent of the 6,000 to 10,000 tonnes caught by small-scale fisheries in 1986 (Michael Batty, personal communication 1986).
Tokelau	On the basis of a household consumption survey on Atafu Atoll, 11 percent of total annual subsistence fish catch was determined to be tuna (Zann and Aleta 1984).
Tuvalu	Of the subsistence catch on Nanumea, 8.9 percent is tunas and other pelagic fish (Zann 1980).
	Skipjack and yellowfin composed 50 percent of the total artisanal fish catch in 1978 (Crossland and Grandperrin 1979).
Wallis/Futuna	Occasional fishing for tuna by only a few residents produces a catch of less than 2 tonnes per year (South Pacific Commission Tuna Program 1984).
Western Samoa	Skipjack, other tuna, mackerel, and barracuda made up 17.5 percent of the total annual fish catch of 1,089 tonnes in 1978 (Western Samoa Department of Statistics 1979).
	Tuna catches of the artisanal fleet increased from 413 tonnes in 1972 to 1,440 tonnes in 1982 (South Pacific Commission Tuna Program 1984a).

lagoon, the reef fish resources are extremely small. It would indeed be a clever writer who could adequately express the jubilation caused by a sailing canoe arriving at Satawal fully laden with over a tonne of tuna. The crew of the canoe pound their paddles with joy while waiting offshore, old women dance and sing on the beach, and the entire population is in a state of delightful anticipation of bone-free protein. Tuna is very important to Satawal.

Recently the use of rafts anchored offshore and of fish-aggregating devices (FADs) has increased the level of small-scale tuna fishing throughout the Pacific. FADs eliminate much of the time, effort, and expense required to locate and chase tuna schools, thereby improving the productivity of fishermen. In some countries—Western Samoa, for example—FADs have been largely responsible for the tremendous increases in the artisanal tuna catch during the past ten years.

CONCLUSION

The results of this field survey demonstrate that tuna is very important to the residents of Fakaofo. The dietary contribution of tuna and tunalike fish—about 16 percent of all animal food consumed during the survey period—was substantial, despite the considerable inflow of money to the island and the consequent availability of imported foods.

Fragmented data from other Pacific islands suggests that the importance of small-scale tuna fishing varies greatly between areas. Tuna appears to dominate local catches on smaller, more isolated islands, especially those with limited resources of inshore fish.

NOTES

1. For descriptions of tuna fisheries see Nordoff 1930 (Society Islands), Kennedy 1930 (Tuvalu), Hornell 1950 (Samoa), Gillett 1985 (Tokelau), Gillett 1987 (Caroline Islands), Hulo n.d. (Papua New Guinea).

2. In this report "tuna and tunalike" refers to those scombrids in the tribes *Thunnini*, *Sardini*, and *Scomberomorini* within the subfamily Scombrini.

3. The remaining 15 percent is used for pig and chicken food.

4. For example, in July and August 1986, $8,952 (NZ$17,552) and $9,659 (NZ$18,939) flowed into Fakaofo from civil service salaries and overseas bank remittances.

REFERENCES

Clark, Les

    1986   Tuna industry developments in the southwest Pacific. In Proceedings of the INFOFISH Tuna Trade Conference. Bangkok. pp. 151–159.

Cook Islands Ministry of Agriculture and Fisheries

    1979   Statistics of artisanal fishing in the southern group 1978. Statistical working paper no. 13. Agriculture Planning Unit. Rarotonga. 13 pp.

Crossland, James, and R. Grandperrin

1979   Fisheries directory of the South Pacific Commission region. South Pacific Commission. Noumea, New Caledonia. pp. 3–33.

Fakahau, Semisi, and M. Shephard

1986   Fisheries research needs in the South Pacific. Forum Fisheries Agency. Honiara, Solomon Islands. 90 pp.

Food and Agriculture Organization (FAO)

1984   Report of the World Conference on Fisheries Management and Development. Rome. 50 pp.

Gillett, Robert

1985   Traditional tuna fishing in Tokelau. Topic review no. 27. South Pacific Regional Environment Program. South Pacific Commission. Noumea, New Caledonia. pp. 1–47.

1987   Traditional tuna fishing: A study at Satawal, central Caroline Islands. Bulletin in Anthropology. Bishop Museum. Honolulu. In press.

Gulbrandsen, O.

1977   Report on travel to Tokelau. Food and Agriculture Organization (FAO) report. Rome. 6 pp.

Hooper, Anthony

1984   Tokelau fishing in traditional and modern contexts. In Proceedings of the UNESCO-ROSTEA Seminar on Traditional Management of Coastal Systems. Edited by K. Ruddle and R. Johannes. Jakarta. pp. 11–12.

Hornell, J.

1950   Fishing in many waters. Cambridge. University Press.

Hulo, John

n.d.   Fishing practices in Buka Island, North Solomons Province. In Subsistence fishing practices of Papua New Guinea. Appropriate Technology Development Institute. Lae, Papua New Guinea. pp. 28–30.

Kennedy, D. G.

1930   Field notes on the culture of Vaitupu, Ellice Islands. Journal of the Polynesian Society 39(1):39–60.

Kiribati Ministry of Commerce and Industry

n.d.   A report on fish catches and fish consumption in the south Tarawa region. Fisheries Division. Tarawa. 79 pp.

Kirifi, Hosea

n.d.   Tokelauan fishing methods. Mimeo. 4 pp.

Macgregor, Gordon

1937   Ethnology of Tokelau Islands. Bulletin 146. Bishop Museum. Honolulu. pp. 108–112.

Mees, C. C.

n.d.   The fisheries of Temana and Arorae. Fisheries Division. Ministry of Natural Resources. Tarawa, Kiribati. Mimeo. 6 pp.

Nordoff, Charles

1930   Notes on the offshore fishing in the Society Islands. Journal of the Polynesian Society 39(2 & 3):1–79.

South Pacific Commission

1984   The tuna program fisheries statistical system. Working paper no. 3. Meeting of coastal states and distant-water fishing nations. Noumea, New Caledonia. 4 pp.

1985   Review of progress, problems, and opportunities within the Tuna and Billfish Assessment Program. Working paper no. 3. Seventeenth Regional Technical Meeting on Fisheries. Noumea, New Caledonia. 2 pp.

South Pacific Commission Tuna Program

1984a   An assessment of the skipjack and baitfish resources of Western Samoa. Final country report no. 14. Skipjack Survey and Assessment Program. South Pacific Commission. Noumea, New Caledonia. 5 pp.

1984b   An assessment of the skipjack and baitfish resources of Wallis and Futuna. Final country report no. 19. Skipjack Survey and Assessment Program. South Pacific Commission. Noumea, New Caledonia. pp. 1–2.

Western Samoa Department of Statistics

1979   Fishery catch assessment survey 1978. Apia. 7 pp.

Zann, Leon

1980   Tuvalu's subsistence fisheries. Institute of Marine Resources. University of the South Pacific. Suva, Fiji.

Zann, Leon, and S. Aleta

1984   A preliminary survey of fish consumption in Tokelau. Institute of Marine Resources. University of the South Pacific. Suva, Fiji.

11.
American Samoa:
The Tuna Industry and the Economy

Donald M. Schug and Alfonso P. Galea'i

INTRODUCTION

American Samoa is an unincorporated territory of the United States with a population of 35,600. For more than three decades the territory's deepwater harbor at Pago Pago has been the site of a successful tuna-processing industry. Since its inception the local tuna industry has been the largest private-sector employer in American Samoa and its leading exporter.

This study first examines how the tuna industry in American Samoa has changed over time in supplies of fish, types of products, and production output.[1] This historical overview is followed by a description of some of the positive and negative economic aspects of the tuna-processing industry in American Samoa. The intent is not to provide a socioeconomic impact assessment of the tuna industry but rather to describe the mixed, often ambivalent, effects of large-scale industrial development on American Samoa's economy. As Miller (1981) notes, the tuna canneries in American Samoa offer an instructive case history in industrial development in a small island-state.

BACKGROUND

The development of the tuna industry in American Samoa began in 1949, when Island Packers, a U.S. corporation financed by the Rockefeller Foundation, installed a small cannery on government property on the north shore of Pago Pago Bay. However, the plant never operated on more than a trial basis because of inadequate supplies of raw tuna; it was not until 1953 that the future of tuna-processing operations in American Samoa was secured. That year the U.S. Bureau of Customs ruled that fish caught by foreign flag vessels could be landed directly in American Samoa. This ruling was crucial to the development of the local industry because the Japanese fishing fleet had dominated the South Pacific tuna fishery up to that time.

A second statutory provision advantageous to the industry in American Samoa was Headnote 3(a) of the U.S. Tariff Schedules. This legislation was critical because the United States is the sole destination of canned tuna produced in American Samoa. According to Headnote 3(a), exports from American Samoa are accorded duty-free entry to the United States if the foreign component value of the product is less than 50 percent of its market value. This criterion is easily met by canned tuna, regardless of the origin of the raw fish. The exemption is especially important for tuna canned in oil, which otherwise is subject to a duty of 35 percent ad valorem. In comparison, the duty on tuna packed in water is 6 percent, increasing to 12.5 percent whenever the quantity of canned tuna imported into the United States exceeds 20 percent of the amount canned in the United States (excluding American Samoa's production) during the preceding calendar year.

In 1954 these statutory provisions induced the Van Camp Seafood Company to lease the site of the defunct Island Packers cannery from the American Samoa government (Figure 1). Processing operations began the following year with fish delivered under contract by Japanese longline vessels. Star-Kist Foods established a cannery in 1963 on leased land adjacent to the Van Camp facility. During that same year Star-Kist Foods became a wholly owned subsidiary of the H. J. Heinz Company, and the Van Camp Sea Food Company was acquired by the Ralston Purina Company. Both parent companies are large, diversified U.S. multinational corporations.

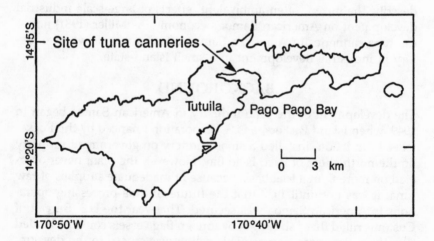

Figure 1. Location of the tuna canneries on the island of Tutuila in American Samoa

After 1963 the tuna industry in American Samoa expanded steadily, in spite of periodic major changes in harvesting and processing. In the mid-1960s Japanese longline vessels began to withdraw from American Samoa in order to fish for Japan's more lucrative sashimi market, and Korean and Taiwanese longline vessels became the tuna-packing companies' major fish suppliers. The longline fleet based in Pago Pago in the early 1970s exceeded 300 vessels. Cannery production faltered between 1975 and 1977 because of reduced landings by the longline fleet, but production increased again after 1978, when the U.S. purse seine fleet shifted from the eastern to the western Pacific. The majority of U.S. purse seine vessels operating in the western Pacific offload in American Samoa; some transshipment takes place in Guam and the Northern Mariana Islands.

Before the longline boats were replaced by purse seine vessels as the principal suppliers to the two canneries, American Samoa's tuna industry specialized in processing albacore (*Thunnus alalunga*) for the premium-priced white-meat canned-tuna market. In 1977 a quarter of the white-meat tuna consumed in the United States was produced in American Samoa (Hester and Broadhead 1980). As purse seine landings of skipjack tuna (*Katsuwonus pelamis*) and yellowfin tuna (*Thunnus albacares*) began to surpass longline landings of albacore, the American Samoa tuna industry assumed a new role as a high-volume processor of light-meat tuna (Table 1).

A second change in product type occurred during the early 1980s. As Table 2 shows, most of the tuna processed in American Samoa during the last few years has been packed in water rather than oil in response to a change in the dietary preference of American consumers.

Table 1. Estimated quantity of fish landed at tuna canneries in American Samoa by species 1981–1985

	Tonnes (000's)				
	Albacore	Yellowfin	Bigeye	Skipjack	Total
1981	20.1	16.9	2.6	23.6	63.2
1982	18.5	10.2	4.5	17.5	50.7
1983	18.9	33.1	4.7	78.8	135.5
1984	10.9	24.0	2.3	63.0	100.2
1985	14.8	24.0	1.7	69.5	110.0

Source: Compiled from unpublished monthly reports between 1981 and 1985 submitted by Star-Kist Samoa and Samoa Packing Company to the Economic Development and Planning Office, American Samoa Government.

Table 2. Estimated quantity of shipments of tuna cannery products from American Samoa to the United States 1977–1985

	Tonnes (000's)			
	Canned tuna in oil	Canned tuna not in oil	Pet food	Fish meal
1977	8.9	7.0	4.1	1.6
1978	17.5	15.0	5.4	1.8
1979	21.9	12.5	5.7	2.1
1980	16.0	20.0	3.2	1.4
1981	15.1	28.8	5.4	1.9
1982	5.9	31.3	5.5	0.9
1983	34.8	42.5	8.1	1.0
1984	19.6	39.9	8.4	2.6
1985	26.7	36.3	11.0	2.8

Sources: U.S. Department of Labor 1986 and U.S. Bureau of the Census 1986.

The change in favor of water-packed tuna weakened American Samoa's advantage of duty-free access to the U.S. market.

While the U.S. purse seine fleet was shifting its efforts to the western Pacific, U.S. processors were closing their operations on the U.S. mainland and expanding their offshore operations in Puerto Rico and American Samoa. American Samoa's tuna-processing industry, in particular, underwent a major expansion because of its proximity to the western Pacific tuna fishery and its low labor costs relative to Puerto Rico's. The percentage of U.S. production attributable to American Samoa canneries increased from 13.2 percent in 1980 to 28.8 percent in 1983 (Table 3). It dropped to 21.3 percent in 1984 but resumed its upward trend in 1985.

ECONOMIC IMPACT

This section provides a brief, largely qualitative overview of the economic impact of the harvesting and processing of tuna on American Samoa in terms of employment and income, skills transfer, interindustry effects, and public-sector revenues and expenditures.

Employment and Income

Because tuna packing is a labor-intensive manual process, canneries in American Samoa have significantly expanded local employment op-

Table 3. Comparison of total U.S. production of canned tuna with American Samoa production 1977–1985

| | Tonnes (000's) | | |
	Total U.S. production	American Samoa production	Percentage of U.S. production
1977	249.1	15.9	6.4
1978	320.4	32.5	10.1
1979	281.9	34.4	12.2
1980	273.6	36.0	13.2
1981	285.0	43.9	15.4
1982	244.8	37.2	15.2
1983	268.4	77.3	28.8
1984	279.2	59.5	21.3
1985	247.7	63.0	25.4

Sources: U.S. Department of Labor 1986 and U.S. National Marine Fisheries Service 1986.

portunities. In 1985 about 27 percent of the employed labor force in the territory was engaged in tuna processing (Table 4). Cannery workers constitute 72 percent of the private-sector employees covered by the statutory minimum-wage level established by the U.S. Department of Labor (U.S. Department of Labor 1986). Periodic increases have raised the minimum wage to $2.82 per hour. Total local wages paid by the two canneries in 1985 were about $15 million.

Labor unions have not played a role in determining the wages and fringe benefits received by cannery workers. In 1975 the U.S. National Labor Relations Board extended its jurisdiction to American Samoa, making it possible for U.S. unions to expand into the territory, but unions have not been successfully established in either cannery except for a two-year period during the late 1970s.

Over the years cannery jobs in American Samoa have increasingly been filled by foreign nationals. Doumenge (1966) estimated that in 1963 about 20 percent of cannery employees were not born in American Samoa. A household survey conducted under the guidance of the South Pacific Commission in 1985 showed that the proportion of cannery workers born outside American Samoa had risen to 70 percent. About 90 percent of the alien workers are Western Samoans who migrate to American Samoa for jobs and higher wages.

American Samoans today generally prefer to seek jobs in local government, in trade and service sectors, or in the United States. To

Table 4. Comparison of total employed labor force in American Samoa
with employment levels of tuna canneries 1977–1985

	Total labor force	Cannery employees	Percentage of total labor force
1977	7,815	1,444	18.5
1978	9,790	1,701	17.4
1979	10,288	2,055	20.0
1980	—	—	—
1981	10,982	2,068	18.8
1982	11,300	2,157	19.1
1983	11,627	2,726	23.4
1984	11,936	2,913	24.4
1985	12,001	3,251	27.1

Source: Compiled from unpublished surveys conducted by the Office of Manpower
Resources, American Samoa Government.

help the canneries meet their employment requirements, the governor
of American Samoa has waived local-hire provisions contained in the
canneries' land-lease agreements and income-tax exemptions. The
American Samoa Immigration Board, in turn, has permitted large numbers of aliens to enter and work in American Samoa.

A consequence of the large proportion of aliens employed in the
canneries is that much of the payroll of the canneries "leaks" out of
the territory. No reliable quantitative data are available, but the strong
family ties characteristic of Samoan culture suggest that Western
Samoans working in American Samoa remit large portions of their earnings to their home country. This leakage of purchasing power from the
local economy diminishes secondary employment and income effects.
Furthermore, because a large percentage of American Samoan males
of working age migrate to the United States for education, military service, and employment (Miller 1981), the majority of secondary jobs
created are filled by aliens.

The presence of the tuna industry has had little beneficial effect
on the commercial fishing activities undertaken by American Samoans.
The domestic fishing fleet consists principally of small vessels with a
limited fishing range. Local small-boat fishermen sometimes sell tuna
to the canneries, but they can usually sell their catches in the local fresh-
fish market for two or three times the price the tuna-packing companies pay. Some local fishermen claim that the tuna industry has, in fact,
reduced the profitability of their fishing operations. They complain that

a portion of the fish landed and frozen by foreign tuna vessels enters the local market, creating an oversupply and depressing prices for fish caught by local fishermen.

Harsh working conditions, low wages, and long fishing trips have discouraged American Samoans from working on foreign longline vessels delivering tuna to the local canneries. They prefer employment on purse seine vessels, but the capital-intensive nature of purse seine operations limits the number of job opportunities. Fewer than a dozen Samoans work aboard the 30 to 40 purse seine vessels currently supplying the canneries.

Skills Transfer

The transfer of job skills to indigenous employees in the canneries has been constrained by the low skill requirements of the majority of cannery jobs and a high turnover in entry-level jobs. Doumenge (1966) suggests that the labor force is unstable because of the nature of canning operations and the characteristics of the workers. The canneries routinely recruit and lay off workers as production levels fluctuate. Employment also tends to be brief because most jobs are nonskilled and low-paying. About 53 percent of the cannery work force consists of unskilled female fish cleaners paid the minimum wage who tend to leave their jobs whenever the family group no longer needs supplementary income. American Samoan cannery workers, moreover, are inclined to leave as soon as they acquire basic skills that can potentially gain them employment in the United States.

The result of this high turnover rate is that few experienced workers are available for promotion to more skilled positions. Managerial jobs in both canneries have consistently been filled primarily by U.S. citizens. It is estimated that fewer than 10 percent of the management positions in the canneries are held by American Samoans.

When the canneries were established, the American Samoa government wanted American Samoans to be trained not only as tuna processors but also as tuna fishermen. In response, the Van Camp cannery purchased a longline vessel in 1961 and operated it with a crew consisting partly of Samoan trainees. The Samoan crew members, however, did not react favorably to the long fishing trips and rigorous working conditions, and Van Camp abandoned the training project after two years. No further major attempts have been made to train Samoans for distant-water fishing.

Interindustry Effects

Both tuna canneries in American Samoa are tied to multinational corporations that supply virtually everything but unskilled labor, shipping services, and infrastructure facilities. A large portion of the raw tuna is now even landed by company-owned vessels. The result is that few backward linkages have developed, and the fish-processing facilities exist essentially as industrial enclaves.

The effects of the tuna-vessel fleet on the local economy are more favorable. Both purse seine and longline vessels refuel, reprovision, and undergo repair and maintenance in Pago Pago. Since 1983 the number of tuna vessels supplying the canneries in American Samoa has declined, but the number of cannery deliveries per vessel has increased. Business representatives say that purse seine vessels based in American Samoa purchase an average of $200,000 in local goods and services per trip. Local purchases by longline vessels average about $72,000 per trip. It is estimated that the tuna fleet spent $29 million for local goods and services in 1985. Servicing of the tuna fleet has both created primary jobs and promoted local entrepreneurs. However, the fleet generates little indirect economic activity because the fuel, food, and other supplies it purchases are almost all imports.

Local investment opportunities have also been created by the availability of miscellaneous fish (fish caught by the tuna vessels that is unsuitable for canning). About 3,500 tonnes of miscellaneous fish—primarily oversize tuna, billfish, and shark—is landed each year. Most of the miscellaneous fish is transshipped by foreign fish traders to Japan for processing in the minced-fish industry, but American Samoan fish dealers are becoming increasingly involved in miscellaneous-fish transactions. Recently a local entrepreneur established a business exporting selected miscellaneous fish to food-service distributors in the United States, and several locally owned businesses export dried shark fins purchased from the tuna vessels.

By accelerating the development of shipping services and port infrastructure, tuna harvesting and processing have fostered the development of other industries in American Samoa, most notably transshipment (Rayacich and Rohlen 1980). The large volume of tuna products exported ensures that an efficient shipping service is available between the United States and American Samoa.

Public Revenues and Costs

Both canneries have received partial or complete corporate income-tax exemptions during much of the time they have been operating in American Samoa. The renewal of tax exemptions has been predicated upon major capital expansion projects undertaken by the canneries. The most recent renewals, granted in 1983 for a ten-year period, obligate the canneries to invest $9 million in capital improvements and to increase employment levels by about 20 percent. These expansions are considered desirable by the canneries to accommodate the increasing supply of tuna from the western Pacific and to compensate for the closing of processing facilities in California. Despite these tax concessions, income taxes paid by the canneries accounted for about 20 percent of the local revenues received by the American Samoa government in 1984. The government also generates revenues from taxes on the incomes of cannery employees and excise taxes on petroleum products purchased by the tuna fleet.

Gains in an expanded tax base have been partly offset by the costs of providing public facilities and services to the industry itself and to an expanding population. Large-scale tuna processing requires a well-developed port infrastructure because of the size of the fishing vessels and the containerized freight facilities used for delivering the product to distant markets. Since 1980 the American Samoa government has spent over $10 million to expand the wharf and container-storage area and rehabilitate the shipyard.

Processing large volumes of fish also requires great quantities of water and electrical power. In 1985 the canneries consumed about 33 percent of the water supplied by the American Samoa government and about 15 percent of the electricity. Because these public utilities are provided to commercial and residential users below cost, they produce a net cash outflow.

Immigration induced by potential employment in the canneries and in secondary activities has increased the territory's overall population, thereby increasing the demand for such government services as education, health care, and law enforcement. Increased income-tax revenues from foreign workers are not sufficient to compensate for increased public-service costs. These services, as well as the public infrastructure facilities, are financed primarily by grants from the U.S. government.

OUTLOOK

American Samoa has attained an important position in the tuna industry in the central and western Pacific, and its status is not likely to change in the near future. The reasons that first attracted tuna processors to American Samoa, such as duty-free access to the U.S. market, still give it an advantage over other production locations in the Pacific region. The tuna industry in American Samoa also enjoys generous tax concessions, government-subsidized infrastructure facilities and services, low fuel prices, and favorable freight rates to the west coast of the United States.

However, the long-term future of the tuna-processing industry in American Samoa must be regarded as uncertain because of the rapid and far-reaching changes occurring in the world tuna industry. The development that could have the greatest negative impact on the territory's tuna-packing operations is the dramatic increase in U.S. imports of canned tuna from Thailand. Although the two tuna processors operating in American Samoa have recently expanded their production facilities in the territory, both also market canned tuna produced by domestic processors in Thailand. Canning operations in countries other than Thailand—Mexico and the Philippines, for example—also have the capacity to develop into important suppliers to the United States. The continuing success of the tuna-processing industry in American Samoa thus depends upon its ability to remain competitive with other low-cost producers.

An event that could have a favorable impact on American Samoa is the expansion of the southern Pacific surface troll fishery into a major new source of albacore. As many as 30 boats from southern California are expected to exploit the fishery in 1987. These trolling vessels are likely to stop in American Samoa for fuel and provisions on their way south from the established North Pacific albacore fishery, thus enhancing the local business activities that now support the purse seine and longline fleets.

CONCLUSION

The number and complexity of issues make it difficult to measure the contribution of tuna harvesting and processing to the economic development of American Samoa. It is certain, however, that loss of the industry would gravely depress America Samoa's employment and income as well as the viability of its fiscal position. American Samoa has essentially a single-industry economy. Canned tuna and related products account for 98 percent of the value of its exports. In short, expansion of American Samoa's economy depends largely upon the growth of processed-fish exports (McPhee 1981). The American Samoa government, well aware of the territory's vulnerability to severe and sudden economic downturns, is stepping up its efforts to promote activities that are independent of the tuna industry.

Beyond endeavoring to broaden the territory's economic base, the government is also taking steps to increase the local economic benefits associated with the tuna industry and mitigate the costs. Many of the negative effects of the tuna industry appear to be consequences of the openness of the American Samoa economy. There are few restrictions on the movement of money, goods, or people (Miller 1981). During the past few years, however, the government has become more involved in the issues associated with the tuna industry. It has encouraged local participation in the purchase, processing, and export of the miscellaneous fish mentioned earlier. It has restructured its immigration policies to limit the number of aliens entering American Samoa. It is imposing stricter quality standards on cannery wastewater discharges. Perhaps most important, the government is monitoring the operations of the tuna canneries more closely. Overall, the trend is toward greater government regulation of the industry to ensure that American Samoa meets its economic development goals.

NOTE

1.   The authors wish to acknowledge the assistance of Jesse M. Floyd in providing background material for the preparation of this paper.

REFERENCES

Doumenge, F.

   1966   The social and economic effects of tuna fishing in the South Pacific. Technical paper no. 149. South Pacific Commission. Noumea, New Caledonia. 42 pp.

Hester, F., and G. Broadhead

   1980   Tuna fishery development plan. Pacific Tuna Development Foundation. Honolulu. 164 pp.

McPhee, M.

   1981   An economic development strategy for American Samoa. Economic Development and Planning Office. American Samoa Government. Pago Pago. 163 pp.

Miller, R.

   1981   The economy of American Samoa. Office of Territorial and International Affairs. U.S. Department of the Interior. Washington, D.C. 68 pp.

Rayacich, D., and T. Rohlen

   1980   Pago Pago commercial port: The potential for transshipment and projected land requirements for the container yard. Economic Development and Planning Office. American Samoa Government. Pago Pago. 62 pp.

U.S. Bureau of the Census

   1986   U.S. trade with Puerto Rico and U.S. possessions. Report FT-800/Annual 1985. Washington, D.C. 106 pp.

U.S. Department of Labor
1986   Various industries in American Samoa. Wage and Hour
            Division. Employment Standards Administration.
            Washington, D.C. 63 pp.

U.S. National Marine Fisheries Service
1986   Fisheries of the United States 1985. Current fishery statis-
            tics no. 8380. Washington, D.C. 121 pp.

12.
High Speed on an Unmade Road: Solomon Islands' Joint-Venture Route to a Tuna Fishery

Anthony V. Hughes

INTRODUCTION

About 1.3 million square kilometers of the western Pacific Ocean is enclosed by Solomon Islands' 200-mile exclusive economic zone (EEZ).[1] The zone is 50 times the size of the nation's land area. The waters in this zone harbor large stocks of tuna, probably migratory species as well as resident species, including the surface-swimming skipjack tuna (*Katsuwonus pelamis*) and small yellowfin (*Thunnus albacares*) and the deep-swimming albacore (*Thunnus alalunga*), bigeye (*Thunnus obesus*), and larger yellowfin tuna (Solomon Islands Government 1985). The fishing is particularly good in the archipelagic waters around the main islands, where rivers, lagoons, and coastal ecology support abundant smaller fish species, including several kinds of anchovies, herring, and sprat on which the surface-swimming tunas feed (Figure 1).

Starting from scratch 16 years ago, Solomon Islands' tuna fishery has grown rapidly to become a major force in today's economy. It provides nearly 7 percent of all formal employment: about 1,850 persons, all but 200 of them Solomon Islanders, work in the fishing industry. The fixed assets of the industry comprise two shore bases, 40 fishing boats and auxiliary vessels (28 of them locally owned), and supporting infrastructure. Two locally incorporated fishing companies, majority-owned by Solomon Islands, catch 35,000 to 40,000 tonnes of skipjack and yellowfin tuna a year. Tuna sales in 1986 were about $30 million, accounting for over 40 percent of Solomon Islands' total exports and a quarter of its cash national product.

The rapid growth of Solomon Islands' tuna industry began when foreign commercial interests and domestic political needs coincided with the existence of an abundant natural resource.

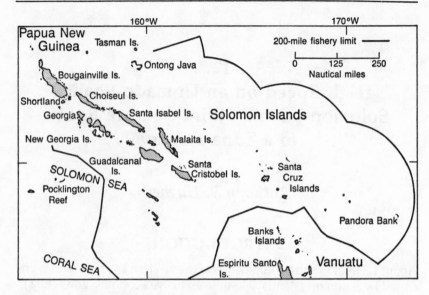

Figure 1. Approximate 200-mile limit of Solomon Islands

ORIGINS OF THE TUNA FISHERY

As late as 1970 there was no hint of what lay ahead in Solomon Islands fisheries. In 1969 a World Bank study of Solomon Islands' economy had merely remarked that "potential for commercial fishing has still to be assessed"; it made no mention of fisheries in its summary of development possibilities (World Bank 1969). The 1971-73 development plan of the then-British Solomon Islands Protectorate had observed that its waters were "rich in fish of many species, but as yet little is known of these . . . resources. . . . There is no large-scale commercial fishing for tuna . . . such as takes place in . . . New Hebrides, Fiji and Samoa" (Solomon Islands Government 1971). The level of government attention paid to fisheries matched the absence of commercial activity. The 1971-73 development plan recorded that the Department of Agriculture "is responsible for fisheries, though it has been inactive for some years in the absence of a Fisheries Officer. Government owns a small fisheries research vessel which has been diverted to other uses."

By a combination of chance and powerful political and economic forces at work elsewhere, this state of ignorance and inaction was blown away. In the late 1960s the colonial administration in Solomon Islands was becoming aware of the need to create a level of economic activity that could reasonably support political independence. Political leaders, looking ahead to budget making in an independent state, shared the concern for finding new sources of revenue. Overseas investments were being sought in tree-crop agriculture and extractive forestry. Some administrators and politicians believed there must also be investment opportunities in fisheries if the resource could be proven productive.

At the same time, sovereign coastal states around the world were asserting extended claims to control of their adjoining seas from 3 miles to 12 miles, then to 200 miles. The Law of the Sea conferences were increasingly pointing to a time when all coastal states would claim 200-mile jurisdiction over their surrounding waters.

The Japanese distant-water fishing industry, assisted by the government of Japan, read the signs early. Japanese longline fleets already ranged the world in search of deep-swimming tunas and were attuned to long-term cost trends and the accessibility of fish stocks. Japan's home-based pole-and-line tuna fleet had expanded beyond the capacity of the bait and skipjack resources that lay within reach to sustain it. In particular, the Taiyo Fishery Company, having grown from a small, family-owned fishing company, was by 1970 a worldwide operator of fishing, marketing, and food-processing businesses. Taiyo had reviewed its strategy for the 1970s in the light of commercial opportunities and political changes; like other Japanese fishing companies, it was looking for overseas bases for its tuna operations. Following a chance meeting with a Solomon Islands official, Taiyo representatives visited Honiara early in 1971 to discuss investment possibilities. Taiyo found a receptive attitude among politicians and officials (James 1979). Events moved quickly. Solomon Islands government agreed to a survey of skipjack and bait resources and trial fishing in 1971-72. The survey would yield data on which investment proposals could be formulated.

Thus, with little assessment of alternatives, Solomon Islands took the first steps down the joint-venture road to developing its fisheries—steps taken with little knowledge of the resource and with almost no government capability in resource management. The knowledge and the capability were to grow as the industry established itself.

The approach was risky, but the imminence of political independence, the apparent dearth of large-scale investment alternatives for generating foreign exchange income, and a feeling that fisheries must be "right" for an island nation in the western Pacific gave the government the courage to proceed. Public criticism within Solomon Islands of the speed and manner of reaching preliminary agreement with Taiyo had a salutary effect upon the government, inducing it to prepare more carefully for the next stage of negotiations.

PROPOSAL AND AGREEMENT FOR A
JOINT-VENTURE FISHING COMPANY (1972–1973)

After a year of successful trial fishing, using at first 4 and later 14 catcher vessels and motherships, Taiyo in 1972 proposed establishing a joint-venture fishing company. The proposal envisaged pole-and-line, long-line, and purse seine fishing with Japanese technology, three shore bases, and an annual catch over 20,000 tonnes. The joint venture would be owned 50 percent by Taiyo, 25 percent by Solomon Islands government, and 25 percent by unspecified Solomon Islands commercial interests. Taiyo would manage the venture, and all vessels would be chartered from Japan. A five-year localization plan was proposed for all but top supervisory staff. Taiyo proposed that it would have exclusive rights to fish commerically for tuna in Solomon Islands waters. Taiyo also sought base facilities for its own distant-water fleet (James 1979).

In preparation for negotiations, Solomon Islands government declared a 12-mile fisheries limit, recruited professional fisheries staff, and consulted widely on the pros and cons of fisheries joint ventures. It engaged experts in law, finance, and fisheries to reinforce its team for the negotiations. The government sought a more distinct and local corporate entity than Taiyo had proposed, with an arms-length relationship with Taiyo and a stronger commitment to shore-base development, including processing. It wanted specific undertakings on localization and a time limit on the exclusivity of fishing rights. The government wanted its 25 percent equity participation to be paid by Taiyo; it also wanted an option to purchase additional equity later. It sought to impose a 10 percent export duty on tuna and insisted there be no base facilities for vessels not part of the joint venture.

Negotiations were conducted, breaking points were reached, trade-offs were made, and an agreement was concluded. The joint-venture agreement (JVA 1), signed in November 1972, provided for a 25 percent "free ride" for Solomon Islands government and a phased introduction of a 10 percent export duty on tuna. Shore bases were to be established following a feasibility study, but it was agreed that the first base would be built immediately at Tulagi. The government imposed an annual catch limit of 30,000 tonnes of skipjack; this limit was set as a conservative estimate of sustainable yield in the main group archipelago. Targets for localization of employment were set in percentage terms for vessels and shore operations.

The venture's governing board consisted of five directors, three from Taiyo and two from Solomon Islands government. The joint venture was granted exclusive rights to fish for skipjack and other tunalike fish within Solomon Islands' waters and to process and export this fish; only Solomon Islanders and "bona fide local companies" would not be affected by this restriction. Taiyo was assigned exclusive rights to market the joint venture's products outside Solomon Islands, on condition that it sell only on the open market at most favorable commercial prices, verifiable by the government. Similar provisions applied to obtaining vessels for the operation. The government retained extensive powers to call for information, to inspect, and to approve the joint venture's contracts with Taiyo. Significantly, purse seining and longlining were specifically excluded from the agreement. The government reserved its position about these methods of fishing.

The financial arrangements allowed both the government and Taiyo to take income off the top of the company's cash flow rather than from its profits. This provision, together with high loan gearing—a 6 to 1 debt-to-equity ratio—made it difficult for the company to accumulate reserves against hard times or apply profits to strengthening itself.

JOINT-VENTURE OPERATIONS (1973–1981)

Backed by excellent survey results, the joint-venture company, Solomon Taiyo, Ltd., (STL) began operations in 1973. The Tulagi cannery was built. It opened in August with a 45-tonne brine freezer tank, a 600-tonne cold store, and a small cannery capable of producing 600 cases of export-grade tuna and 300 to 350 cases a day of dark-meat dry-pack tuna. The cannery, established at low cost with mostly secondhand equipment, quickly became a financial success.

With the pole-and-line fishery established, Taiyo in 1973 obtained approval from Solomon Islands government to survey the nation's longlining potential. The survey was interrupted by logistical problems in 1974–75, but it resumed in 1976. The government was keen to see a domestic longline fishery established before declaring extended jurisdiction, which it planned to do in 1977 or 1978. After substantial "trial" fishing and the accumulation of two years of data, however, Solomon Islands government and Taiyo could not agree on terms for a permanent, locally based longline operation because of an economic problem. The distance from the Japanese market, rising production costs, and falling catch rates were making longlining an unprofitable method of fishing for the cannery market. And entering Japan's lucrative sashimi market required a costly investment in superfreezing vessels.[2] Failure to reach agreement with Taiyo on domestic longlining operations disappointed the government and influenced its subsequent decisions on purse seining.

By mid-1975, following a feasibility study and Solomon Islands government agreement, the second base was under construction at Noro. It was opened for the 1976 season. The base at Noro was larger and better laid out than the one at Tulagi, having an 80-tonne brine freezer and 800 tonnes of cold storage. With the opening of the Noro base and an expansion of fishing operations, employment of Solomon Islands nationals by STL exceeded 600 (James 1979). The catch had expanded as expected, passing 15,000 tonnes in 1976 and contributing more than a third of the country's total export earnings. The company was profitable after paying export taxes to Solomon Islands government and sales comisssions to Taiyo, but these offtakes by the shareholders precluded the payment of dividends or accumulation of reserves. Of its $9 million 1976 sales revenues, $3.8 million (42 percent) was paid to Okinawan vessel owners and crews operating under a charter agreement. This was a fairly typical ratio during the 1970s.[3]

Further important changes were in the making in the latter 1970s. Taiyo had made changes in its top management in 1976–77, and its corporate strategy was undergoing further review. The idea of more shore bases for STL was quietly dropped, the government itself recognizing the need to consolidate the gains so rapidly achieved. Expansion to the remote Santa Cruz group (the eastern extremity of Solomon Islands) was discussed, but no commitment was made. The government threatened to open up that area to other transnational exploitation, but Taiyo made only token efforts to exercise its rights there. A marked lull in development set in.

Solomon Islands, having become internally self-governing in 1976, was clearly set for political independence from Britain in 1978. Independence was achieved in an atmosphere of stability and steady economic growth, but foreign investors had adopted a cautious attitude during the preparatory period.

STL's first four years had shown the government that it was unrealistic to expect the joint company to take what appeared to its management to be imprudent (and for Taiyo unrewarding) commercial risks in localizing senior employment or vessel ownership. Nor were Taiyo's managers in STL likely to have much time for training their own replacements; they were fully occupied in running the company. It would also, in the view of Solomon Islands officials, be unwise to get to the end of JVA 1 with no alternative vehicle available for tuna industry development if a follow-on joint venture could not be agreed. Thus the government took the opportunity, in the context of negotiations in 1977–78 about a supplementary agreement on a longlining pilot project, to tighten the provisions in JVA 1 about crew localization and to relax the requirement that STL expand its own fleet. It also announced its intention to establish a second tuna fishing company, National Fisheries Development, Ltd. (NFD).

Originally conceived as a way of accelerating local ownership of the fleet to facilitate duty-free access to European markets, NFD was quickly perceived also as a potential vehicle for developing local boat-building capacity, providing more intensive training of fishing vessel crews, and speeding up development of national management capability in the tuna industry. The Asian Development Bank (ADB) embraced the concept enthusiastically, seeing it as a way to increase lending both to the fisheries sector and to the Pacific islands and as a counter to what it felt was excessive dependence on Japanese technical support in fisheries development.

The concept in 1977 was for ten ferro-cement pole-and-line boats to be constructed in Solomon Islands by NFD, with associated fiberglass bait boats. NFD would operate, initially at least, as a supplier of fish to STL and a training ground for crews, officers, and company management to accelerate localization throughout the industry, including STL. An existing boatyard at Tulagi was acquired, expanded, and re-equipped. Two pole-and-line vessels were supplied to Solomon Islands government under Japanese aid in 1979. The government made them available on charter to NFD.

The creation of NFD in 1978 usefully brought a third force into the production side of the tuna industry alongside STL's own boats and the Okinawan chartered vessels. And by introducing some competition into the industry, the new company stimulated better personnel practices and took some of the localization pressure off STL.

During the 1978-79 period the Forum Fisheries Agency (FFA) was being established in Honiara. Solomon Islands proclaimed a 200-mile exclusive economic zone in January 1978 and opened negotiations for access to the zone by Japanese distant-water tuna fleets. Negotiations with the seasoned warriors of Japan's fishery authorities led eventually to licensing agreements. These were subsequently renegotiated annually.[4] With few exceptions, Solomon Islands had very satisfactory experiences with the Japanese authorities in provision of data and compliance with agreements, setting standards of performance that other fishing nations were unable or unwilling to achieve.[5] Solomon Islands, like other island countries, looked forward to the day when a larger share of the value of the tuna harvested by distant-water fleets in its waters could be extracted by coordinated action by island countries; hence Solomon Islands was a keen supporter of the move to establish FFA.

There was a strong interaction between Solomon Islands' knowledge of and attitude to distant-water fishing in the 200-mile zone and its experience with the domestic tuna fishery. The same government officials were negotiating access agreements as were monitoring the local pole-and-line operations and trying to establish a domestic longline fishery, and the interplay of experience was invaluable. The Fisheries Division of the Ministry of Natural Resources developed confidence and authority, publishing in 1978 a management plan for the nation's tuna resources that set benchmarks for resource utilization on a conservative basis, providing a framework both for access agreements and for expansion of the domestic industry (Solomon Islands Government 1978).

By the end of 1979, NFD was fishing with its two Japanese aid boats, and its first two ferro-cement boats were under construction. Remedial action had been taken in 1979 to put STL on a firmer financial footing. Solomon Islands government then proposed to Taiyo a second joint-venture agreement, indicating the broad lines on which it might be drawn up. The government wanted to see further development of fleet, shore bases, and processing; it wanted an increased role for Solomon Islands in ownership and management of STL (including participation in marketing), improved training, and higher standards of personnel management; it wanted Taiyo to continue its roles in investment, financing, management, and sales. Taiyo, however, preferred to make as few changes as possible, apart from the already-agreed fleet and cannery developments. Solomon Islands government assembled a group of specialists to assist the home team, and a round of negotiations began that took almost two years to conclude.

PROBLEMS IN THE FISHERY

By 1974 Solomon Islands government recognized that it needed an adequate management capability for its tuna resource. It established a fisheries division in the Ministry of Natural Resources, recruited additional professional staff, and arranged for closer monitoring of baitfishing and tuna catching and marketing. The remedies were due.

The government's fisheries officer later described the situation at the time as "chaotic," with "local opinion and Government officials hardening their attitude towards [STL]. . . . This was the end of the honeymoon period. . . . Many of the problems stemmed from cultural friction between the [STL] expatriate staff and the local population" (James 1979). There were mutual failures in communication: Japanese staff made inadequate efforts to understand Solomon Islands culture, feelings, and patterns of thought; Solomon Islanders held unrealistic expectations and low commitment. Relations improved somewhat in the mid-1970s when Solomon Islands officials became more active in managing and directing STL. Localization of crews was accelerated, putting foreign crews in the minority. This action not only increased the confidence and productivity of local crew members but also reduced wage costs. Accommodations ashore were improved. Formal agreements were negotiated with customary owners of baitfishing grounds for regular payments of royalties. These payments went to a locally controlled community fund; in some cases they went to individual "chiefs" who distributed them among the community.

This period was one of rapid expansion for STL, and many important details had to be left for later attention. Most of the problems concerned people rather than fish. It has been well argued that little thought had been given to social and cultural factors surrounding the creation of an industrialized fishery, and little or no study had been made of factors contributing to transient work patterns and low levels of productivity of Solomon Islands employees (Meltzoff 1980).

It is remarkable that more open and bitter differences did not flare up between local and foreign workers in the tuna industry and between industry employees and villagers. There were occasional fights and at least one murder for employment-related reasons, but there was no sustained pattern of open confrontation between ethnic or linguistic groups.[6] The relative smoothness of relationships can probably be attributed to the management skills of a few key STL employers who were able to spot and deal with potential troubles and to the later establishment of improved personnel procedures and better education of the Japanese managers in STL.[7]

The joint venture's other problems were financial ones. The original provisions permitting a high ratio of debt to equity and allowing Taiyo and the government to take income out of cash flow rather than out of profits had an unfortunate result: it allowed the shareholders to become insensitive to deterioration in the company's earnings position; so long as commissions and taxes were being paid, neither shareholder was motivated to move quickly to avert losses. As a consequence, the company several times had to turn to its shareholders for "rescue" operations after its financial position had been seriously undermined by low catch rates, external cost hikes, or adverse market conditions.

When Solomon Islands government was deciding to establish NFD in 1978, an adverse combination of catch rates, prices, and exchange rate movements brought the shareholders of STL together to consider the company's financial position. Catches were poor locally in the early part of the year, but overseas markets were well supplied and prices had fallen. Meanwhile the yen had appreciated from an average of 350 to one Solomon Islands dollar in 1976 to 230 to the dollar in 1978. The yen-denominated fish price used as the charter fee for the Okinawan catch was now above the price STL could get for the fish, and the company was incurring substantial losses. Additional exchange losses were incurred on yen-denominated debt. Too highly geared to withstand a marked drop in earnings and lacking financial reserves, the company turned to its shareholders, and Taiyo turned to Solomon Islands government.[8]

After heavy negotiations in Honiara and Tokyo in late 1978 and early 1979, remedial measures were agreed on. These included (1) provision of additional share capital by both partners, in the course of which the Solomon Islands shareholder moved to a 49 percent position; (2) reduction in the payments for Okinawan vessels and crews; (3) ordering of three new catcher boats by STL (not unconnected with Solomon Islands government's move to establish the second fishing company, NFD); (4) commitments to detailed planning of a new cannery to be built at Noro; (5) increased use of Solomon Islands dollar loan finance to reduce future exchange risk; and (5) an additional Solomon Islands director on the board, now to be three from each side, with a Taiyo chairman. Once they got going, these negotiations were carried on by both sides with a sense of urgency and purpose. The exchanges gave Solomon Islands government valuable insights into Taiyo's philosophy and method of operation, which assisted it in negotiating a second joint-venture agreement.

SECOND JOINT-VENTURE AGREEMENT (1979–1981)

The second joint-venture agreement (JVA 2), signed in November 1981 with a year of JVA 1 still to run, provided for equality of shareholding for five years and thereafter 51 percent for the Solomon Islander shareholder. Each partner would appoint three directors. The board chairman would be a Solomon Islanders' shareholder's nominee but would have no vote. The general manager would be a Taiyo nominee for the duration of the agreement, but the finance manager would cease to be a Taiyo nominee after five years. A personnel and training department and a marketing department would be established, each headed by a board appointee.

The second JVA included more detailed provisions than its predecessor on information disclosure, access to company documents, marketing, local and overseas procurement, and monitoring and reporting of intercompany transactions and relationships. STL retained its exclusive fishing and processing rights (except for bona fide local companies) in the "archipelagic waters and territorial seas" (that is, from a baseline around the archipelago plus 12 miles). These rights were conditional upon STL's paying a reasonable commercial price and actually exercising the rights. (STL lost these rights in 1982 in the Santa Cruz area because it did not use them.) Taiyo would be exclusively contracted by STL for overseas sales except for the canned product in the Pacific islands area. The commitment to build the Noro cannery was made explicit, as were undertakings to provide services and amenities broadly commensurate with good citizenship. Export duties were put on a sliding scale, and the joint company was to pay import duty on fuel, from which it had been exempted under JVA 1. The effect of the financial arrangements was, again, to provide the shareholders and Solomon Islands government with substantial benefits before striking a profit, leaving little prospect for the company to build financial reserves. JVA 2 was to run for 11 years, to 1992.

JOINT-VENTURE OPERATIONS (1981–1986)

The JVA 2 negotiations had taken place against a background of rising tuna prices and optimism about future price levels; the agreement was signed amid anticipations of strong earnings and early investment in new assets. Increasing attention was being paid worldwide to apparently large stocks of easily accessible skipjack and surface-swimming yellowfin, estimated in the western Pacific alone to comprise 3 to 4 million tonnes, of which only about 600,000 tonnes was being harvested. Vast cost-driven structural changes were afoot in the tuna industry as high U.S. labor costs first slashed margins, then forced the closing of U.S. domestic canneries and their relocation to Puerto Rico and American Samoa.

Meanwhile, other developing countries were moving to establish tuna fisheries aimed at the same markets for frozen and canned fish as STL, and the U.S. tuna fleet re-equipped with long-range boats to follow the skipjack and yellowfin schools westward across the Pacific. Both Solomon Islands and Taiyo had been affected by price euphoria in negotiating JVA 2. The lessons of 1978 had been buried under STL's good financial results in 1979–81. But the financial projections underlying the fiscal and commercial aspects of JVA 2 were quickly undermined as tuna prices fell in the face of sluggish demand and rising supply through 1982 and the next three years.

Serious financial problems began to mount. Both STL and NFD incurred operating losses and accumulated substantial short- and long-term debt, requiring financial assistance from their shareholders to remain in business. Faced with lengthening delays in provision of port and town infrastructure for the Noro cannery, STL squeezed 25 percent additional product from the existing cannery at Tulagi by installing some machinery destined in due course for Noro.

By now the tuna industry was firmly established in the economy and the community of Solomon Islands. Pole-and-line catches, which had been around 25,000 tonnes in 1979–81, dipped to 19,000 tonnes in 1982, then rose to 30,000 tonnes in 1983 and 1984. Almost 40 percent of the country's total export earnings was coming from tuna. As each NFD ferro-cement boat joined the fleet, hopes for effective localization grew; STL's new vessels, built in Japan, were likewise greeted as evidence of a strong industry. In 1982 Solomon Islands government undertook to provide the port and town infrastructure at Noro that would support the planned cannery, and official Japanese sources agreed to provide long-term, soft-loan financing for the cannery itself. Planning by the relevant government agencies got slowly under way in 1983, and in 1984 the government requested funding from the European Development Fund under the Lome Convention. This was a frustratingly slow process for STL, which had had plans and financing for the cannery ready since 1982 and was now seeing its financial position slip back each year from its high point in 1980–81.

In 1984 the expansion of the U.S. fleet into the western Pacific, which was known to have led to many unlicensed intrusions into Solomon Islands waters, finally produced a long-awaited confrontation. An American purse seiner, the *Jeannette Diana*, was reported by an STL pole-and-line boat to be fishing well inside the 200-mile limit. Arrested next day by a government patrol boat and brought to Honiara, the owner and master were found guilty of unlicensed fishing. They did their cause little good by deriding the High Court of Solomon Islands as a kangaroo court. The vessel was confiscated, and fines were imposed.

The United States responded with an embargo on Solomon Islands fish exports to the United States, acting under domestic legislation designed to protect U.S. fishermen from arrests such as this. The political outcry that followed the U.S. action brought about an unusual degree of cohesion among Pacific island countries. The FFA was now able to orchestrate concerted demands that the U.S. government agree to negotiate a multilateral access treaty for American purse seiners.

The U.S. embargo was an unwelcome burden to Solomon Islands' tuna industry, but it was not a crippling financial blow. Prices were already depressed. Fish intended for sale to Puerto Rico and American Samoa was sold, after some price discounting, on other markets, notably Thailand (whence much of it found its way in cans to the United States). The embargo was lifted early in 1985, when Solomon Islands sold the vessel back to its owners for $700,000.[9]

In 1984–85 Solomon Islands, through its shareholding agency, undertook to split NFD into two separate companies, one retaining the shipbuilding and ship-repair functions and assets, the other taking the fishing task and vessels. In 1986 the shipyard, now called Sasape Marina, took over the government's marine fleet base at Tulagi; it is now poised for further capital investment and a new strategic plan. The fishing company, now bearing the NFD name, is also set for major developments, again assisted by the Asian Development Bank, with establishment of its fleet management at Tulagi and the acquisition, by charter, of two 500-GRT purse seiners now under construction in Australia.

A BACKWARD LOOK

From a vantage point of 15 years after the events that prompted Solomon Islands to join forces with Taiyo in a tuna fishery, one can assess the progress of the joint venture.

The initial survey of skipjack and bait resources and the year of trial fishing in 1971–72 proved to be good indicators of what was possible. For example, the trial fishing for surface-swimming tuna yielded an average catch per fishing day of 5.8 tonnes. Though this record has never been equaled, averages of over 4 tonnes per day have been recorded in 8 of the last 16 years, and the cumulative average is 4 tonnes. Table 1 shows catch and effort for the 15 years to 1985.

Table 1. Summary of pole-and-line catch and effort 1971–1985

	Effort			Catch (tonnes)		
	Boat-months	Total fishing days	Mean fishing days per month	Total	Av. catch per fishing day	Av. catch per boat-month
1971	34	813	24	4,711	5.80	138.57
1972	145	3,356	23	7,905	2.36	54.52
1973	81	1,944	24	6,512	3.35	80.40
1974	92	2,182	24	10,331	4.73	112.30
1975	108	2,419	22	7,169	2.96	66.38
1976	145	3,495	24	15,799	4.52	108.96
1977	197	4,741	24	12,115	2.56	61.50
1978	187	4,656	25	17,454	3.75	93.34
1979	207	5,085	25	23,800	4.68	114.98
1980	201	4,993	25	21,935	4.39	109.13
1981	219	5,259	24	22,626	4.30	103.32
1982	197	4,858	25	17,322	3.57	87.93
1983	264	6,185	23	29,266	4.73	110.86
1984	260	6,397	25	30,599	4.78	117.69
1985	318	6,966	22	25,234	3.62	79.36

Source: Solomon Islands Government 1986.

The skipjack and yellowfin resources of the main group archipelago, the fishing ground of STL and NFD, show no signs of depletion. The 1985 annual report of the Fisheries Division cited estimates of sustainable yields of 83,500 tonnes of all tunas from Solomon Islands waters (Solomon Islands Government 1986). This overall estimate had been geographically allocated in a fisheries policy paper published the previous year (Solomon Islands government 1985) indicating total allowable catch limitations as follows:

Pole-and-line and purse seine fishery	Main group archipelago	40,000 tonnes
	Temotu archipelago (Santa Cruz group)	15,000 tonnes
	Other areas within fishery limits	20,000 tonnes
Total		75,000 tonnes
Longline fishery, total all areas		8,500 tonnes

The longlining limit was set below the calculated maximum sustainable yield to allow room for development of a locally based longline fishery. This had earlier received a setback when arrangements to integrate the two 80–GRT longliners operated by NFD into a larger operation involving refrigerated transport to Japan could not be brought to fruition. (The two vessels, after four years of technically successful but commercially loss-making operations, were withdrawn from fishing early in 1986. No longliners are currently fishing the deep-swimming tunas within Solomon Islands' EEZ.)

As for NFD, the second company, initial projections for the build-up of the catch rate were overly optimistic. However, as the company progressively took delivery of its own ferro-cement boats from 1980 to 1985, its catches have risen to over 20 percent of pole-and-line production, worth $5 million in 1986. NFD sells all of its catch to STL, which in turn is obliged to buy it at a reasonable commercial price, construed in recent years to be the same as that paid for fish supplied by Okinawan vessels. Solomon Islands government and Taiyo have now agreed that from 1989 on, NFD will be able to decide its own marketing arrangements on stand-alone commercial criteria.

Bait for the pole-and-line fishery has continued to be readily available except for localized shortages, usually attributable to self-correcting conditions such as sediment from rain-swollen rivers clouding the lagoon bait grounds. (Pressure on baitfish stocks is monitored by the Fisheries Division, and bait grounds may be closed if there are signs of stress.) Bait catches, 260 tonnes in 1973, the first year of STL's operations, rose to 2,234 tonnes in 1985. There was no sign of generalized overfishing of the bait stocks (Solomon Islands government 1986), but there has been a marked rise in the 1980s in the amount of bait needed to catch each tonne of tuna. Table 2 gives details of the bait catch for the 13 years of STL's existence.

Financial difficulties continue to plague STL. The tuna industry's problem of depressed market prices was compounded for STL in 1985–86 by the appreciation of the Japanese yen. The exchange rate went from 186 yen to one Solomon Islands dollar at the start of 1985 to 80 yen at the end of 1986. The extent of the yen's appreciation surprised most people, apparently including Taiyo. The structure of STL's finances was such that the company recorded substantial exchange losses, and the effect of high gearing and lack of internal reserves was once more brought home to the shareholders. Measures recently agreed provide for reductions in shareholder offtakes and yen-denominated operating costs, an increase in equity, and conversion of substantial yen-denominated short-term debt to Solomon Islands-dollar long-term liabilities.

Table 2. Annual baitfish catch and effort for the locally based pole-and-line fleet in Solomon Islands 1973–1985

	Effort	Catch			
		Bait	Buckets	Tonnes	
	Nights fished	catch (buckets)[a]	per night	Bait catch	Bait:tuna ratio
1973	1,722	118,808	69.0	261.4	1:24.9
1974	1,503	91,371	60.8	201.0	1:51.4
1975	1,563	130,587	83.5	287.3	1:24.9
1976	1,967	157,685	85.2	368.9	1:42.9
1977	2,913	225,076	77.3	495.2	1:24.4
1978	3,597	238,965	66.4	525.7	1:33.2
1979	4,858	303,741	62.5	668.2	1:35.6
1980	4,903	325,645	66.4	716.4	1:30.6
1981	4,892	645,811	132.0	1,420.8	1:15.9
1982	5,335	672,203	126.0	1,478.8	1:11.7
1983	6,844	895,631	130.9	1,970.4	1:14.8
1984	6,548	813,570	124.2	1,789.8	1:17.1
1985	7,593	1,015,539	133.8	2,234.2	1:11.3

Source: Solomon Islands Government 1986.
[a]One bucket = 2.2 kilograms.

The resulting STL financial structure is expected to provide a sound basis both for viable operations and for the planned investment in processing at Noro. The European Development Fund-financed Noro infrastructure project has now reached the implementation stage. The shareholders of STL have confirmed the corrected timetable for the construction of the Noro cannery in 1987–88 and put in hand the raising of the necessary investment funds.

As for NFD, it is obvious now that the costs for building its boats were underestimated and the speed with which catch rates would build up was overestimated. Probably too little account was taken of the costs of training and working up human skills and productivity. As a result, substantial additional capital has had to be injected to support both the building program and fleet operations. The necessary commercial realism has not come easily to NFD, but the company can now point to solid achievements in the development of boat-building skills, self-confidence in employees, and the making of a fishing company from rather unpromising circumstances.

NFD is now undergoing major changes to enable it to broaden its operational base and prepare to receive the purse seiners under construction and being brought in on charter. The operation of these vessels requires considerable expansion of NFD's management and support capacity. Solomon Islands government and Taiyo have agreed to modify the terms of their joint-venture agreement to provide for NFD to take over the Tulagi base in 1989, while STL will concentrate its fishing and processing operations at Noro. These moves imply further changes to NFD's assets, liabilities, and manpower as this company expands the scope of its operations.

Efforts to achieve localization in Solomon Islands' tuna industry are gradually showing results. Fishing crews are slowly becoming more professional and permanent employees of the industry, returning for successive seasons and accumulating expertise that shows up in productivity. Most fishermen have little or no secondary education, and though new intakes are from the upper levels of secondary schooling, by 1986 only 72 of the industry's 1,450 national employees had formal deck or engineering qualifications. After more than ten years of operations, the manning ratios of the most productive pole-and-line boats still comprised 7 or 8 Okinawans and about 25 Solomon Islanders. The aim of creating a professionally skilled and productive pool of fishing-boat crews has to be painstakingly achieved by more formal training, active personnel management, and rewards for service.

NFD uses a handful of expatriate personnel in management, advisory, and line positions; it employs Okinawan fishermen in a limited number of key positions on about half its fishing vessels, the others being fully localized. Over 90 percent of its 400 employees are Solomon Islanders. Replacement of foreign crews as well as fishing vessels has been a prime target for NFD. Considerable progress has been made, despite some well-publicized setbacks. Invaluable knowledge of what makes a fishing crew of Solomon Islanders "tick" is being accumulated in NFD. More localized ashore and afloat than STL, it has the opportunity to break even at the lower catch rates of its smaller, ferro-cement boats and to be profitable at lower fish prices than STL and the Okinawan boats. Where NFD goes in localization, STL is bound to follow, once the productivity problem is solved, for good commercial reasons.

Production in Solomon Islands' tuna industry has continued to expand in spite of the considerable financial strain it has been under since 1982. Following delivery of the three new boats ordered in 1979, STL has bought four more boats secondhand from Okinawan owners. NFD took delivery of its last ferro-cement boat in 1985. The 1986 pole-and-line fleet comprises 10 vessels owned and operated by STL, 12 by NFD, and 12 owned by Okinawan companies and operated by STL under a charter-fishing agreement. Of the total of 34 vessels, 15 are typically based at Tulagi and 19 at Noro. Seven of the vessels freeze the catch on board; the others chill or ice the fish for freezing at the shore base.

The pole-and-line fleet is now thought to be about its optimum size, at least until a more detailed scientific study has been made of the bait resources available to support it. One boat from Tuvalu is now fishing with the fleet, and a Solomon Islands privately owned pole-and-line boat is expected to join the fleet in 1987.

Potential appears to exist for increased production by purse seining within the total allowable catch set by the government (Solomon Islands Government 1986). For many years the government has held a cautious policy toward this efficient but drastic method of catching surface-swimming schools of skipjack and yellowfin. The only purse seine operation presently allowed in Solomon Islands waters is a group seine operation by STL. Using a 180–GRT net boat, a small scout/support vessel, and two refrigerated fish carriers (converted distant-water pole-and-line vessels), this unit fishes mainly around fish-aggregating devices (FADs) moored just outside the 12-mile limit.

Begun as a pilot operation in 1980, the unit is now owned and commercially operated by STL. Excellent catches have been achieved, averaging well over 400 tonnes a month and exceeding 8,000 tonnes in a 12-month operation during 1985–86. Very high productivity such as this enables the purse seine operation to afford the high cost of key Japanese personnel on board and still make a profit.

The government has now decided on a rapid expansion of locally based purse seining and has invited both STL and NFD to step up their operations by the use of chartered vessels under the flag of Solomon Islands. STL has responded cautiously, aware that the extent of interaction between purse seining and pole-and-line fishery is not yet established, with a proposal to add one single-vessel purse seiner to its existing operation. NFD, with a smaller pole-and-line operation, is moving to introduce a number of chartered purse seiners in 1987. The government has also placed orders for two new 500-GRT purse seiners for delivery in 1988. Capital costs of these vessels, under construction in Australia, are being met by a blend of official Australian aid and export credits. Modern vessels in all respects, these new ships will place considerable demands on management and crews, and Solomon Islands is arranging for financial and technical assistance from the Asian Development Bank for the initial years of operation of the vessels by NFD.

CONCLUSIONS

Some conclusions stand out clearly 15 years after Solomon Islands set out on the joint-venture road to a tuna industry. It is unlikely that the industry could have been established so fast, on such a scale, and in such a permanent form by any method but a joint venture. Both shareholders in STL have achieved a substantial part of their initial aims, divergent though those aims undoubtedly were and still are. There has been sufficient overlap between the objectives and capabilities of Solomon Islands and those of Taiyo to establish and maintain a joint venture at the core of the tuna industry. Both have met disappointments, and some hard lessons have been handed out by the joint venture to its owners. But the two sides have been intelligent enough to maintain sufficient mutual benefit and interest in continuing the venture and investing in increased processing at Noro.

The financial benefits have been substantial. In the 14 years of STL operations, over 300,000 tonnes of tuna has passed through the company's hands, producing sales revenues of about $223 million. STL has made a profit in 6 of its 14 years. It has paid about 9 percent of its gross revenues to Solomon Islands in taxes, about 5 percent to Taiyo in commissions and fees of various kinds. These cash flows, totaling around $32 million, may be regarded in one sense as an ample return for investment by the shareholders of equity that reached only $1.8 million in 1979 and $3.6 million in 1981. Each of the joint venturers can compare these revenues with what they would have received from other methods of using the same resources—for example, by licensing foreign-based vessels. Both Solomon Islands and Taiyo also receive important indirect benefits—economic and strategic spin-offs for Solomon Islands, for Taiyo a valuable contribution to overheads and wider commercial benefits in its overall corporate development. These benefits are partly offset by some indirect costs for each partner in the form of forgone options in this and related fields.

Taiyo's contribution to the joint venture has been in the industrial know-how and the access to capital and to markets that Solomon Islands sought from its foreign partner. The cost of these has been acceptable to Solomon Islands government, despite recurrent concerns, especially during JVA 1, about the possibility of transfer pricing within Taiyo's connected operations. With changing relative prices (caused in part by the vast appreciation in the yen exchange rate), tighter margins, and increasing competition, STL has to obtain its material and human inputs from the most cost-effective sources available. This requires adjustments by Taiyo itself, as a main supplier and procurer of such inputs under the joint-venture arrangements, with its prices and purchasing arrangements subject to Solomon Islands scrutiny.

On its part, Solomon Islands has made the tuna resource available with sufficient exclusivity to protect the joint venture while actively promoting the growth of NFD as a second national tuna enterprise. The government has several times shown its readiness to modify its taxation offtake in the face of commercial reality. The second joint-venture agreement has provided the government with excellent access to operational and financial data of all kinds, especially hard market information almost impossible to obtain in other ways. It has also learned that, even with a well-drawn agreement, effective monitoring and use of information requires sustained effort by officials and board members and the use of well-prepared periodic meetings between shareholders to review the condition and performance of their joint venture. The government's resource managers and other officials, while enforcing the relevant laws and regulations, have treated the commercial operation reasonably. It appears from recent events that Solomon Islands and Taiyo are determined to place the industry on a sound financial footing to sustain and expand its operations. It remains to be seen if the financial lessons of 1978, repeated in 1985–86, have indeed now been learned.

If commercial success does not come easily, social success is equally elusive. The Solomon Islands' tuna industry reached high levels of production quickly through extensive use of imported vessels and key crew members. The resulting foreign exchange costs are increasingly out of line with the sharper market environment of the 1980s. But neither capital nor manpower can be immediately localized. Costs must be reduced and productivity increased without disrupting the operation. This places a high premium on the management of people.

THE ROAD AHEAD

Human resources rather than fish, boats, or shore bases are likely to be the limiting factor, and increased attention will have to focus on training, motivation, retention, and efficiency. The physical structure of the industry seems set for the next few years while the new cannery is being established and purse seining is being developed alongside the pole-and-line fishery. The roles of STL and NFD, complementing and competing with each other within the overall industry framework, seem now to have been defined for the rest of the 1980s. Increased production and processing will demand increased skill and energy in marketing, with scope for new products to be established and new markets opened up. Looking ahead to the 1990s, further development of the production, processing, and marketing functions and realignment of the associated assets and liabilities should make for more efficient use by the industry of both local and foreign resources.

Choosing the joint venture road 15 years ago has brought the Solomon Islands' tuna industry a long way in a short time, but it has been a bumpy ride. More skillful driving, with a growing ability to handle the controls and effect repairs on the move, should mean fewer breakdowns or upsets. But it seems unlikely that the road itself will get any smoother, and heavier traffic will place additional demands on the drivers.

NOTES

1. The views expressed in the paper are those of the author and do not necessarily represent Solomon Islands government or other agencies concerned with the country's tuna fishery.

2. Ironically, the problem was exacerbated in 1981 when Japan, as part of its aid program, supplied two 80-GRT longliners that were equipped for superfreezing but could not operate economically without support from a mothership. These vessels were delivered after Taiyo and Solomon Islands government had already failed to agree on how such support could be provided commercially.

3. As the locally owned fleet has increased in recent years, the proportion of sales revenue paid for use of foreign-owned vessels and crew has fallen to 25 to 30 percent.

4. In those early days there was much argument about the quality of each side's statistics. On one occasion Solomon Islands government had arrested inside the 12-mile limit a Japanese vessel licensed to fish between the 12- and 200-mile lines. The vessel's logbook recorded a false position. When challenged, the captain said this was a common practice among Japanese fishing vessels; asked why,

he explained that it was to conceal matters from the home authorities. This revelation came conveniently during a phase of negotiations when Solomon Islands was questioning Japanese catch-location data.

5. Solomon Islands government revenue from these licensing agreements approximated 3 to 4 percent of the estimated value of the fish caught, amounting to several hundred thousand dollars per year. Catches by Japanese longliners in Solomon Islands waters averaged 3,270 tonnes annually from 1979 to 1985; the two locally based 80-GRT vessels caught 200 to 500 tonnes a year (Solomon Islands Government 1986). The Japanese long-range pole-and-line fleet reported catches of 43,000 tonnes in the four years 1974–77 within what became Solomon Islands' 200-mile zone in 1978 but very little in subsequent years until 1985, when 3,300 tonnes was reported.

6. Baitfishing brings pole-and-line crews into direct contact with villagers. Bait-boat crewmen, Solomon Islanders or Okinawan Japanese, live ashore at the bait grounds, preparing for night-time baiting by the incoming catcher-boats. Economic and social interchange can create tensions and produce side-effects. Officials of central and provincial governments have had to play a facilitating role in negotiating bait fees and smoothing occasional upsets, which they have done with some success.

7. Working conditions and standards have also improved. In the 1970s there were several eye injuries on the pole-and-line boats from flying hooks, some of which led to legal action for compensation from STL. Fishing masters and STL itself were accused of callous indifference to on-the-job risks. But such cases are almost unheard of in the 1980s. Early in 1986 a crewman died of an arm injury on a purse seiner, but the accident seems to have been the result of a rare individual lapse of caution, since safety standards on the vessel are high.

8. In retrospect, it is clear that STL management and both shareholders were slow to appreciate and react to the nature of the company's problems. There were shades here of things to come seven years later in 1985–86, but apparently few lessons were stored in corporate memories.

9. Resale of the boat was a necessary precondition for access treaty negotiations between the South Pacific coastal states and the United States. These have now been successfully concluded, just two years later.

REFERENCES

James, R. H.

1979 Case study of a fisheries joint venture 1971–1979. FAO technical report no. W/P5265. Rome. 64 pp.

Meltzoff, Sarah, and Li Puna

1980 A Japanese fishing joint venture: Worker experience and national development in Solomon Islands. ICLARM technical report no. 12. International Center for Living Aquatic Resources Management. Manila. 63 pp.

Solomon Islands Government

1971 Sixth development plan 1971–73. Government of the British Solomon Islands Protectorate. Honiara.

1978 Fisheries management plan for billfish, tuna, and tuna-like fish resource of Solomon Islands. Fisheries Division. Ministry of Natural Resources. Honiara.

1985 Policy paper for proposed development of skipjack and tuna fishing industry. Fisheries Division. Ministry of Natural Resources. Honiara.

1986 Fisheries division annual report 1985. Ministry of Natural Resources. Honiara.

World Bank

1969 The economy of the British Solomon Islands Protectorate. International Bank for Reconstruction and Development. Washington, D.C.

13.
Growth and Contraction of Domestic Fisheries: Hawaii's Tuna Industry in the 1980s

Linda Lucas Hudgins and Samuel G. Pooley

INTRODUCTION

Hawaii's commercial fishing industry is undergoing a rapid transition. New tuna fisheries are developing while the old leader, the skipjack tuna (*Katsuwonus pelamis*) fishery, has declined. The skipjack industry lost its international market with the closing of the Castle and Cooke cannery in 1984; at the same time its domestic market share shrank in response to competition from other local tuna fisheries. Furthermore, availability of the skipjack tuna resource around Hawaii apparently declined and fishing costs increased, further exacerbating problems for the industry.

This paper examines events that are restructuring the fishing industry in Hawaii. Particular attention is given to the effects on the skipjack industry of (1) the closing of the tuna cannery, (2) the changed availability of skipjack around Hawaii, (3) the increasing costs of fishing for skipjack, and (4) competition from other tuna species in the Hawaii fresh-fish market. The first three sections of the paper describe recent tuna fisheries developments in Hawaii. A section on marketing tuna in Hawaii follows. The last section discusses policy issues relevant to expansion of domestic tuna production.

SKIPJACK TUNA INDUSTRY AND CANNERY OPERATIONS

For almost 70 years the skipjack tuna industry was Hawaii's major commercial fishery, supplying 68 percent of the state's annual landings (over 4,000 tonnes on long-run average), 42 percent of its annual ex-vessel revenues ($4 million in 1986 prices), and almost 100 percent of its seafood exports.[1] In recent years annual skipjack production declined to approximately 1,300 tonnes, worth $2.7 million. This decline in total catches is shown in Figure 1. Pole-and-line skipjack now constitutes

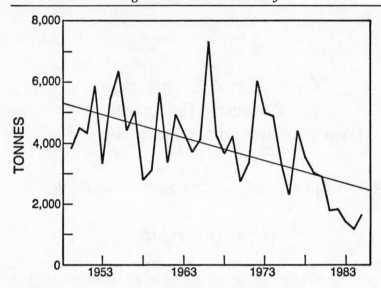

Figure 1. Skipjack tuna catch around Hawaii 1948–1983

Source: Skillman 1987.

about 22 percent of all tuna catches in Hawaii, down from a long-run average of 80 percent.

This fishery, known locally as the *aku* fishery, has existed since at least the turn of this century. Before World War II there were as many as 26 wooden-hulled vessels employing 300 fishermen using pole-and-line technology to fish with live bait for skipjack tuna. Over the years the skipjack fishermen have adopted only such minor technological advances as partial conversion to fiberglass poles. The vessels have crews of eight to ten and must forage for bait in Hawaii's harbors and bays before each trip. Today only eight vessels are actively fishing; one larger steel-hulled vessel is tied up waiting to be sold. Until 1984 the vessels sold both to the fresh market and to the tuna cannery; now they sell their catches entirely on the domestic fresh-fish market.

F. Walter Macfarland, a pineapple planter, founded Hawaii's tuna cannery, Hawaiian Tuna Packers (HTP), in 1917. He brought fishermen from Japan, built two gasoline-powered sampans, and signed five of the then-existing sailing sampans to supply the cannery with tuna. This facility packed up to 100,000 cases by 1922 (Castle and Cooke, Inc. 1977).

The existing four-line cannery building was constructed in 1932 at Kewalo Basin, the major commercial fishing harbor in Honolulu. The U.S. military took over the entire fishing fleet during World War II and converted the cannery to production of airplane gas tanks. After World War II HTP purchased about half of all skipjack catches from the remaining bait-boat fleet (17 vessels) and imported frozen tuna to supplement these domestic catches, processing about 72 tonnes of canned tuna per eight-hour day. HTP marketed the canned product under the Bumble Bee and Coral labels. The cannery also turned out Figaro cat food, fishmeal, and solubles used as additives for cattle and poultry feed. It was the site of the only commercial ice plant on Oahu, with a production capacity of 77 tonnes of ice per 24-hour day and a cold storage facility with a capacity of 1,760 tonnes. HTP also managed a shipyard on its Kewalo Basin premises.

The cannery changed ownership several times from 1917 onward. Castle and Cooke acquired it after World War II and owned it in one form or another until closing it down in 1984 because of corporate cash flow difficulties.[2] In 1985 Castle and Cooke sold the cannery, without the shipyard, to a group of Hawaii investors, the WRAF Corporation, who intend to integrate the cannery into a marine-oriented tourist center.[3] Under conditions of the 40-year land lease with the State of Hawaii, the WRAF Corporation must use the premises for a "minimum one-line fish cannery and the manufacture, storage, and sale of ice. Other permitted uses include the sale of seafood products, beverages and fishing equipment, a seafood restaurant, and supporting facilities. The lessee must also purchase all excess tuna as defined in and required by the lease for the purpose of assisting the local fishing industry" (Honolulu Star-Bulletin 1986). Rental for the land is $111,600 for 1987 and $125,550 for 1988. For each of the next ten years the rent will be $242,000 or 6 percent of gross receipts, whichever is greater. The lease also requires an expenditure of at least $300,000 in capital improvements at the site.

Reopening the cannery, even on a smaller scale, is an important economic and political issue in Hawaii for several reasons. First, the cannery employed over 400 persons. Second, the loss of cannery sales cost the skipjack fleet at least $500,000 in annual sales. Third, the skipjack fleet receives lower prices for its fresh-market sales and has reduced its production levels in response to the loss of the cannery market. Last, loss of an outlet for surplus skipjack tuna destabilized prices on the domestic fresh-fish market. (These cross-price effects are discussed below in the section on marketing.)

Even if the cannery reopens, the Hawaii fleet in its current status is unlikely to be able to catch enough tuna to fully supply the cannery. A one-line cannery requires about 18 tonnes a day (4,000 tonnes a year) to operate at full capacity. The average skipjack vessel in this fleet historically caught about 275 tonnes per year. At this level of productivity, the maximum the nine-vessel fleet could produce is about 2,475 tonnes a year. However, these vessels would sell first to the fresh market (historically around 1,045 tonnes per year), leaving about 1,300 tonnes to sell to the cannery.[4] Using only the locally caught skipjack, the cannery would be operating at about 35 percent of capacity. It could, however, draw on frozen imports, summer surpluses of domestically caught yellowfin tuna (*Thunnus albacares*), and landings from the distant-water albacore (*Thunnus alalunga*) fleet for supplemental supply.

On the other hand, reopening the cannery could attract new investment into Hawaii's skipjack fleet, either traditional pole-and-line boats or vessels with new technology. If these new entrants expect to be able to sell on the domestic market, their catches would have to maintain the high standard of skipjack available through the existing pole-and-line technology. Any new vessel entering the skipjack industry would also face competition in the fresh market from other Hawaii tuna fisheries, which have grown rapidly over the past ten years. In summary then, closing the cannery forced investors in the skipjack industry to consider processing techniques and market channels that had been postponed by reliance on the cannery. If the cannery reopens, some of this investment may yet be undertaken and lead to increased production.

Long before the cannery closed, there was an apparent change in resource availability of skipjack tuna around Hawaii. During the same period, the 1970s, industry production costs also increased. Therefore the closing of the cannery in 1984 may have been only the culminating event that threw the industry into decline. These resource and cost issues are discussed in the next section.

SKIPJACK RESOURCE CHANGES AND PRODUCTION COST INCREASES

Since 1974 the size distribution of catches has shifted from larger skipjack to smaller skipjack, with no documented change in fishing patterns. Large-size tuna as a percentage of total catches dropped from 64 percent in the 1964–73 period to 36 percent in the 1974–82 period (Table 1). Because large fish command a higher price in both Honolulu's fresh markets and cannery markets, the changed size distribution directly affects fishermen's revenues. The change in size of fish caught is estimated to have cost the industry about $362,000 per year in sales (Hudgins 1986). The causes of this change in size distribution are not yet clear. It may be that fishing pressure elsewhere in the Pacific by purse seine vessels and/or in the Northwestern Hawaiian Islands by Japanese pole-and-line vessels has decreased the availability of skipjack to Hawaii fishermen or that biological factors have affected fishing productivity.[5]

Table 1. Average annual size distribution of skipjack tuna catches around Hawaii 1964–1982

| Period | Percentages | | | |
	Large[a]	Medium	Small	Extra small
1964–73	64.1	16.1	15.5	4.1
1974–82	36.3	23.2	28.3	12.0

Source: Hudgins 1986, 12.
[a]Large = over 6.8 kilograms.
Medium = 3.6 to 6.8 kilograms.
Small = 1.8 to 3.6 kilograms.
Extra small = under 1.8 kilograms.

Along with the changed size distribution within catches came a decline in total tonnes caught, resulting in a loss of an additional $1 million per year in sales. For example, during the peak fishing years 1970–78 skipjack catches averaged over 3,200 tonnes a year; by the 1980s total skipjack production had declined to 1,300 tonnes annually. This decline in catches can be attributed primarily to the changing availability of skipjack around Hawaii rather than to cost increases or attrition in the fleet (Hudgins 1986).

Production cost increases, however, also have reduced long-run earnings and profitability in the skipjack industry. Pole-and-line skipjack fishing in Hawaii is relatively high in cost because of such fixed expenses as insurance and a high labor component compared to, for example, purse seine operations elsewhere in the Pacific. By 1983 total costs for an average vessel in the fleet exceeded the revenues as catches declined. With no profits to invest, the fleet's capital base depreciated. As fixed input prices for repairs, insurance, and loanable funds rose during the 1970s, the residual crew share also declined. A rise of 95 percent in inflation-adjusted prices for skipjack tuna during the 1970s may have been the only factor permitting some vessels to continue operating in the face of increasing costs and declining catches.

All but one of the existing vessels are at least 40 years old and require regular maintenance, without which they cannot get insurance. (Insurance is required for entering U.S. government harbor areas to gather baitfish; thus inability to get insurance can put a vessel out of the fishery.) These types of fixed costs are about 20 percent of total costs. The cost picture is further complicated by the fact that the larger vessels have operating-cost problems whereas the smaller vessels have fixed-cost problems. If fuel costs, about 22 percent of total costs, continue to decline, the larger and/or better maintained vessels will benefit. Insurance will, however, continue to be a problem as rates continue to increase (Pooley 1987).

OTHER TUNA FISHERIES

While the skipjack vessels have been affected by the closing of the cannery and by rising costs and resource problems, the other tuna fisheries in Hawaii have begun to grow and become competitors in the fresh-fish market. Estimates of the production of these other tuna fisheries are shown in Figure 2 and Table 2. State of Hawaii statistics report that these other fisheries comprise 55 percent of commercial tuna landings, substantially greater than their share before 1979.

These fisheries are most easily classified by gear type: the longline fishery for yellowfin, bigeye (*Thunnus obesus*), and albacore tunas and other large pelagics; the small-scale handline and troll fisheries for yellowfin and bigeye tunas; and the distant-water fishery for albacore tuna. There is also a large recreational tuna fishery for which few data are available.[6]

The longline fishery for yellowfin and bigeye tuna, and more recently albacore tuna, was traditionally Hawaii's second-largest fishery, highly prized for sashimi. The number of longline vessels declined in the 1960s and 1970s but began to recover by 1980. The number of long-

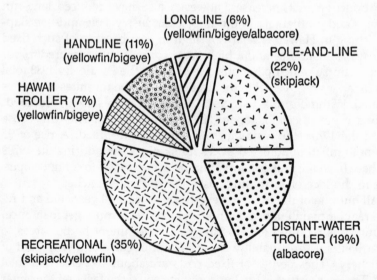

Figure 2. Estimated Hawaii tuna catches by fishery mid-1980s

Adapted from Table 2.

Table 2. Estimate of Hawaii tuna fishery production mid-1980s

Fishery*	Catch in tonnes	Value $m
Pole-and-line (skipjack)[a]	1,300	2.7
Longline (yellowfin, bigeye, albacore)[b]	370	1.7
Handline (yellowfin, bigeye)[c]	615	2.1
Hawaii troller (yellowfin, bigeye)[d]	410	1.1
Subtotal domestic commercial tuna landings	2,695	7.5
Recreational (skipjack, yellowfin)[e]	2,050	5.3
Distant-water troller (albacore)[f]	1,100	1.1
Total tuna landings	5,845	13.9

*This table shows *unofficial* estimates, presented to indicate relative magnitudes of production in Hawaii's tuna fishery. Because of inconsistencies in data reporting, the figures are estimated from the most accurate available sources, which are identified for each fishery. In some cases, only estimates based on knowledge of current operating characteristics could be made. Actual Hawaii Division of Aquatic Resources (HDAR) commercial tuna landings for 1984 were 2,745 tonnes valued at $6.5 million.

[a]Data are from HDAR 1985 monthly summaries (preliminary) and 1984 data by gear type.

[b]Data are extrapolated from HDAR 1984 total tuna landings for longline vessels, and reflect 2.86 X more boats reporting.

[c]Data are HDAR 1984 handline tuna totals for gear types 3, 8, and 9.

[d]Data are HDAR 1984 totals for tuna by gear type 6.

[e]Data are from unpublished National Marine Fisheries Service (NMFS) Marine Recreational Fishing Statistical Survey (MRFSS) results for 1981. Values are based on average commercial troll tuna prices in 1984. The MRFSS data have not been verified and are subject to significant variation.

[f]Data are based on one Hawaii trip per vessel, 60 short tons (54 tonnes) per trip, 20 vessels in the 1986 season. Values are current Honolulu delivery prices, which include transshipment discount.

liners operating in Hawaii today is difficult to determine. However, it is possible that the fleet has tripled since 1975 to approximately 40 vessels in 1986 (Honda 1987). There are also several multipurpose vessels that use longline gear during periods of peak tuna demand. Total catch for all these longline vessels could be as much as 370 tonnes per year ($1.7 million). These landings are marketed primarily on the Honolulu fresh-fish market and on the U.S. mainland. Bicatches of mahimahi (*Coryphaena hippurus, C. equiselis*), wahoo (*Acanthocybium solandri*), and various billfishes are sold in fresh-fish markets throughout Hawaii.

The fastest-growing fishery in Hawaii has been the tuna handline fishery (*palu ahi* and *ika shibi*) on the island of Hawaii. Using very small vessels (10 meters) and traditional fishing techniques, the fishery began with night fishing in the early 1970s. It grew to at least 200 vessels by 1981, then declined between 1981 and 1985. These small vessels can be very profitable because they have low operating and fixed expenses and because opportunity costs for labor are frequently low. This fishery is especially subject to fluctuations in the near-shore availability of tuna and to weather conditions that affect launching, recovery, and fishing effectiveness.

Annual catches of tuna—about 90 percent yellowfin—in the handline fishery could exceed 615 tonnes ($2.1 million). Most of the tuna is sold at the Hilo auction on the island of Hawaii, but some is shipped directly to the Honolulu auction or sold under contract for resale in Honolulu or on the U.S. mainland. Marketing has been constrained by a "burnt tuna" problem, which discolors the meat and substantially reduces its value. Nonetheless, handline-caught yellowfin, bigeye, and albacore tuna have become major competitors with skipjack in Hawaii's fresh-fish market.

There are also commercial and charter boat fleets that troll for tuna and sell on the domestic market; some recreationally caught tuna is also sold. Hawaii's commercial troll fishery supplies approximately 410 tonnes of tuna annually ($1.1 million).

The combined annual catches of Hawaii's non-skipjack tuna fleets, exclusive of the recreational catch, total approximately 1,400 tonnes ($4.9 million). Up to 25 percent of this catch may be exported directly to the U.S. mainland. The remainder—a substantial increase over the 1960s and 1970s—is sold on the fresh-fish market in Hawaii. Recreational catches, possibly the largest of all the fisheries, may exceed 2,000 tonnes annually (estimated value $5.3 million), but this is poorly accounted.[7]

Finally, a part of the U.S. distant-water albacore trolling fleet began using Hawaii as a seasonal base in the late 1970s. By 1982 some 75 distant-water albacore vessels (average size 20 meters catching about 40 tonnes per trip) were fishing the grounds above Midway Islands at the northern end of the uninhabited Northwestern Hawaiian Islands. Many of these vessels transited through Honolulu; some offloaded catch. However, with the decline of world tuna prices, the high salt content of brine-frozen products, and the closing of the Honolulu cannery, the cost of fishing in the central North Pacific became prohibitive for many vessels. Fewer than 25 vessels, 4 of them based in Hawaii, fished these waters in 1986. Total U.S. albacore catches in the central North Pacific in the mid-1980s (not all landed directly in Hawaii) were approximately 1,100 tonnes, valued at $1.1 million. New techniques, including flash-freezing, are being developed by some vessels to enhance fleet revenues. The albacore fleet is not expected to have an important impact on tuna marketing in Hawaii in the future, though it may provide an indirect benefit if vessels home-port in Honolulu or stop over for supplies. However, to the extent that some of these vessels make landings for the domestic fresh-fish market, there may be a substantial impact on bottomfish or pelagic markets.

As these other tuna fisheries developed, marketing practices became more sophisticated, extending outside Hawaii. Although no reliable estimates are available, it is probable that the growth of these fisheries, especially the handline fishery for yellowfin, affects prices and sales of skipjack in the market. Arrangements for marketing tuna in Hawaii are described in the next section.

TUNA MARKETING STRUCTURE

Annual seafood purchases of all species by Hawaii retailers for resale to the public are estimated to be $110 million (23,600 tonnes), with the population of Hawaii consuming twice the national average of seafood (Hudgins 1980b). Figure 3 indicates the source of Hawaii's retail seafood supply.

Tuna caught by Hawaii fishing vessels is about 2,800 tonnes (12 percent) of Hawaii retail sales. Hawaii vessels catching other pelagics and bottomfish provide about 5,900 tonnes (25 percent) to the retail sector

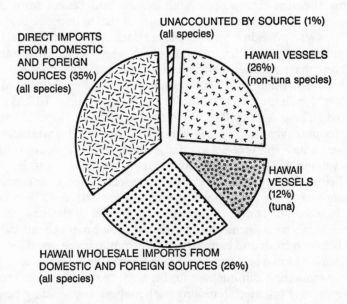

Figure 3. Hawaii retail seafood supply by source, percentage of total purchases (23,050 tonnes) 1983

Sources: Cooper and Pooley 1982; Higuchi and Pooley 1985.

for resale. Wholesalers provide about 6,100 tonnes ($23 million), 26 percent, from domestic and foreign overseas sources to local retailers for resale. The rest of the fish and seafood consumed in Hawaii, about 34 percent valued at $30 million, is imported directly from mainland U.S. suppliers or from foreign suppliers, with a small percentage coming from unspecified local sources (Cooper and Pooley 1982; Higuchi and Pooley 1985).

Much of this marketing is done through bilateral arrangements between wholesaler and retailer and is therefore difficult to examine directly. There are, however, two auctions and two fishermen's cooperatives that are central to marketing tuna in Hawaii. One auction, located near downtown Honolulu on Oahu, is expected to move to the new cannery complex; the other is in Hilo on the island of Hawaii. These auctions are markets for fresh large-size tuna and for bottom-fish. (Only small amounts of skipjack pass through the auctions because skipjack is marketed primarily through the fishermen's cooperatives.) Wholesale dealers are the center of the auction system, which is based on the Japanese model of sale by individual large tuna and by individual lot of smaller fish. Both auction organizations have adapted to the times, welcoming a wider number of local buyers, assisting them in transactions with buyers and sellers from outside Hawaii, and promoting restaurant sales of fish to improve prices. The auctions also provide a number of functions to the commercial fishery (including short-term financing in some cases) which insure that the auctions maintain their centrality to the overall fish market.

Two fishermen's cooperatives have historically marketed skipjack tuna in Honolulu: the Tuna Boat Owners Association (TBOA) and the United Fishing Agency (UFA), which also runs the Honolulu auction. The cooperatives traditionally compete to supply the wholesale tuna market, which then resells to retail outlets. TBOA manages only skipjack vessels; UFA manages both skipjack and other vessels.

Before the cannery closed, TBOA vessels pooled their catches and determined shares of total sales revenues by formula. Each vessel's share was based on its contribution (weighted by size of fish caught) to the cooperative's total catch. The cooperative's sales manager allocated the fish between fresh and cannery markets to maximize revenues. When the cannery closed and membership in the cooperative declined, TBOA's role in marketing diminished, and the cooperative became more administratively oriented, handling such matters as providing health insurance and coordinating fuel and ice purchases for members. The strength of the cooperative was its ability to withhold a certain portion of the catch from the fresh market and sell it to the cannery, recognizing that other independent sellers of skipjack would be price followers. This action resulted in relatively high prices on the fresh market and higher revenues for the skipjack fleet as a whole. It was in the interest of both larger and smaller skipjack vessels to market cooperatively in order to maintain prices over time, as shown in Figure 4. The high-producing vessels, in particular, were able to make catches up to their capacity under the cooperative marketing arrangement and still capture a portion of the higher fresh-market price benefit.

As the range of tuna species available on the local market increased, marketing and market shares have become more important. Today, with the cannery closed, roughly 25 percent of total tuna catches are sold through the local fresh-fish auctions. Another 45 percent of the tuna catch is marketed by the cooperatives. The rest (30 percent) is marketed through contract arrangements between individual vessels and local retail outlets (restaurants and small grocery stores).

Price instability for all species in the fresh-fish market since 1984 is said to be a direct result of excess summer skipjack being put on the market during June, July, and August in competition with auction-sold yellowfin, bigeye, and albacore tuna. Reducing skipjack sales to the fresh market and selling the excess to the cannery would maintain prices on the fresh market. Price maintenance and an alternative sales outlet are especially critical in summer, when the large-size skipjack are caught. If the cannery reopens, a possibility to enhance revenues for the current skipjack fleet will exist in addition to more stability overall for fresh-fish sales.

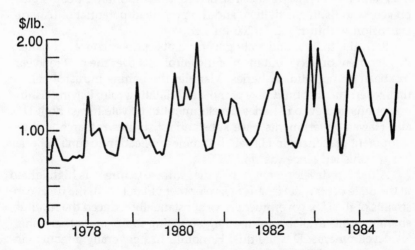

Figure 4. Monthly Hawaii fresh tuna market price trends 1977–1984

Sources: Hudgins (1980a) and unpublished industry data.

Sales of fresh tuna have become more fully integrated into one wholesale market for all tuna in contrast to the segmented markets that existed when the cannery provided an outlet for skipjack. Historically, there has been little observable cross-elasticity between skipjack prices and other tuna prices because the markets were not integrated. (Stronger cross-elasticities imply that the price of one tuna species is affected by the quantity sold of another.) With the restructuring of the skipjack industry so that more skipjack is sold in the fresh market, it appears that these cross-elasticities among tuna species have become stronger. The cross-price effects in the future may radiate throughout the entire fresh fish industry, resulting in losses to other than skipjack producers. Without the cannery sales for the pole-and-line skipjack fleet, all fresh-fish producers in Hawaii will face increased competition until the market stabilizes over the longer term.

INSTITUTIONAL CONSIDERATIONS

The commercial fishing industry in Hawaii is a small contributor to the state's overall economy, similar in scale to other products in Hawaii's diversified agriculture sector. Table 3 shows that locally caught seafood comprises as little as 0.1 percent of direct income to the gross state product, although the overall fishing industry, including a cannery, contributes more through processing, transshipment, and wholesale/retail activities. Nonetheless, fishing is notable for the extent of its integration into Hawaii's social structure, the importance people place on access to fishing activities, and the perceived potential of fisheries expansion within the U.S. 200-mile zone.

Both the federal and state governments have encouraged fishery development projects, but most of the emphasis over the past ten years has been on non-tuna fisheries. Many in the fishing industry believe that government should play at most a facilitating role, leaving fishing operations and even fisheries development to private firms. Both U.S. and Hawaii governments have substantial fisheries research and development presences in Hawaii, but their application to tuna fisheries has diminished since the mid-1970s.

Although development of fisheries infrastructure was highlighted in the State of Hawaii Fisheries Development Plan (1979), the fiscal constraints of the U.S. government have substantially reduced the revenue-sharing grants available for harbor, launching ramp, and other fisheries developments. Despite this, Honolulu has generally adequate infrastructure facilities, and the increased movement of transient fishing vessels through the area—especially for the lobster, albacore tuna, and bottomfish fisheries—has increased the level of support available to

Table 3. Value of production in selected Hawaii industries 1983

	Value in $m	% of total value[a]
Sugar (including processing)	410	3.00
Pineapple (including processing)	219	2.00
Defense	1,848	13.00
Tourism	3,974	28.00
Flowers and nursery products	36	0.30
Aquaculture (raw product)	1	0.01
Commercial fishing (raw product)	10	0.07
Other agriculture (raw product)	115	0.80
Other primary industry and all secondary	7,739	54.00
Gross state product	14,352	100.00

Source: State of Hawaii 1985.
[a]Percentages do not add to 100.0 due to rounding.

local fleets. On the other hand, the limited access to launching ramps for small-scale commercial fishing boats and the unavailability of ice are significant constraints for fisheries development on the neighbor islands.

Other tuna industry problems include seafood quality and baitfish availability. The state's Seafood Product Promotion Committee, which has codified seafood availability, handling practices, and nomenclatures, has played a major role in improving tuna quality. Although considerable funds have been spent on baitfish research, including aquaculture facilities, only a pilot project on the island of Maui shows any near-term promise, and it may be terminated for lack of federal funding.

PERSPECTIVES ON THE TUNA INDUSTRY

The tuna industry in Hawaii is in a transition marked by changing production levels and developing market channels. The skipjack tuna industry has apparently suffered production declines because of the reduced availability of the resource and the closing of the cannery. At the same time, Hawaii's other tuna fisheries have expanded alternative markets for their fresh and frozen product, in particular through vigorous development of overseas markets. This expansion of market channels has had a salutary effect on the entire wholesale seafood industry in Hawaii and has assisted the development of the longline and handline tuna industries and the handline bottomfish industry.

However, the further development of domestic and overseas markets depends on three factors: the growth of the tourist market for seafood in Hawaii (6 million visitors per year), the increased availability of air cargo transportation between Hawaii and the overseas markets, and the modernization of marketing practices by Hawaii seafood dealers.

Despite the complexities and the transitional nature of these developments in the state's tuna fisheries, there is an important conceptual issue to be seen in Hawaii's case. That is, in order to increase production, a domestic fishery will at some point look to external market sales.[8] The development of canneries or sizable transshipment facilities may develop external sales and maximize value added from locally produced tuna, but risks attend such a development pattern. The external market (in Hawaii's case, the tuna cannery) appears to provide a stable outlet for surplus production. However, the external market itself is subject to international competitive price pressures and, in the case of a non-locally owned cannery, sales to the external market may depend on corporate decisions unrelated to the domestic industry. Conditions in the domestic industry and external dependence on the cannery both contributed to the transformation of the Hawaii skipjack industry in the 1980s.

If a fishery begins production on a small scale and expands to industrial scale to supply an external market, the two production levels are linked by the scale of the fishing fleet and by market interactions felt primarily in the domestic seafood sector. Industrial development may not involve a smooth transition from one level of production to another. Production at the output level required for an international commercial market implies more capital investment and higher fixed costs and therefore less flexibility in the face of changing production or market conditions. To amortize this larger investment, the fishing fleet will need to produce at near-full capacity and have a reliable sales outlet for its catch. Large-scale investment undertaken in response to *potential* sales in the externally linked market could lead to collapse of the industry if it must rely solely on domestic sales. In this event the higher-cost vessels would become marginally profitable while their catch of larger volumes of fish would jeopardize the economic well-being of the smaller vessels in the domestic fleet.

An important corollary issue obvious from the Hawaii skipjack industry case is the consideration of not only the degree of capitalization of the fleet relative to available markets but also the type or scale of capitalization. The skipjack industry in Hawaii, although contracted, will survive, precisely because it is based on small-scale technology. When the Hawaii cannery closed, the skipjack industry could either have moved to a lower long-run level of production or collapsed completely. The actual outcome depended on the type of technology adopted by the Hawaii industry over the years. As it was, the pole-and-line technology is relatively small in scale, and although there will be some attrition if the cannery remains closed, a portion of the fleet will remain to supply the domestic market for fresh skipjack whether the cannery reopens or not.[9] The remaining vessels will be those with relatively lower costs. If the technology had been larger in scale, and therefore dependent upon cannery sales, the cannery's closing might have brought about the complete failure of the skipjack fleet in Hawaii. Had the fleet had distant-water capabilities, the cannery's closing might have led to its deployment elsewhere in the Pacific.

It is difficult to assess the relative impacts of the many events that led to the restructuring of the fishing industry in Hawaii: closure of the cannery, decreased availability of the resource, increased production costs, and competition in the market from other tuna fisheries. The future of the skipjack industry is clearly tied to developing new market outlets for surplus production, maintaining fresh-market revenues, and containing costs. The future of the whole tuna industry in Hawaii will depend on continuing to pursue new markets as well as fulfilling current market demands with high-quality fish.

NOTES

1. Data are from unpublished monthly summary reports on commercial landings from the Hawaii Division of Aquatic Resources (HDAR).

2. In 1956 Bumble Bee Seafoods traded 12 percent corporate interest to Castle and Cooke in exchange for the cannery. Castle and Cooke again acquired direct ownership of the cannery in a corporate merger with Bumble Bee Seafoods in 1961 and continued operations until 1984.

3. The four owners of WRAF Corporation are Rex Y. Matsuno (Suisan Co., Ltd., of Hilo, Hawaii), two principals of the United Fishing Agency, Ltd., of Honolulu, and William R. Zappas, a shopping center developer from California.

4. Historical prices would be maintained at about 1,045 tonnes per year sold to the fresh market and 1,550 tonnes allocated to an alternative market. A fleet of 17 vessels of the current average size would sustain both a one-line cannery and the fresh skipjack market using only local production.

5. Boggs and Pooley (1987) review current problems in the skipjack industry with respect to both biology and economics.

6. As many as 15 percent of the Hawaii population engage in the small-boat recreational fishery. Most of the recreational catch sold goes directly to small stores and outlets rather than to established markets and so is not accounted for in official surveys or records.

7. Hawaii Division of Aquatic Resources (HDAR) data on commercial landings have been hampered by incomplete reporting of some large tuna vessels and by the lack of a reporting requirement for the recreational category.

8. In the Hawaii case several species are marketed on the U.S. mainland, which is analogous to an external market linkage.

9. The actual skipjack fleet without cannery sales will most likely be four to five vessels. The outcome depends on the skipjack fleet's competitive position relative to other tuna fleets in the domestic fresh market (Hudgins 1987).

REFERENCES

Boggs, Christofer H., and Samuel G. Pooley.

1987   Forces of change in Hawaii's aku industry 1986: A summary of a workshop. NOAA technical memorandum. National Marine Fisheries Service. Honolulu. In press.

Castle and Cooke, Inc.

1977   Special to Pacific Blue Water Magazine. Press release dated June 24, 1977. Honolulu. 6 pp.

Cooper, James C., and Samuel G. Pooley

1982   Total seafood volume in Hawaii's wholesale fish markets. Administrative report H–82–15. Southwest Fisheries Center. U.S. Department of Commerce. Honolulu. 12 pp.

Higuchi, Wesley K., and Samuel G. Pooley

1985   Hawaii's retail seafood volume. Administrative report H–85–6. Southwest Fisheries Center. U.S. Department of Commerce. Honolulu. 16 pp.

Honda, Victor A.

1987   An updated description of the Hawaii tuna longline fisheries. Southwest Fisheries Center. Honolulu. Mimeo.

Honolulu Star-Bulletin

1986   Notice of public auction. Auction sale of Department of Transportation lease. Legal notice. September 29. p. C–8.

Hudgins, Linda L.

1980a   An econometric model of a fishing market with endogenous supply: The Hawaii skipjack tuna case. Ph.D. diss. University of Hawaii. 114 pp.

1980b   Per capita annual utilization and consumption of fish and shellfish in Hawaii 1970-77. Marine Fisheries Review 42(2):16-20.

1986   Economic issues of the size distribution of fish caught in the Hawaiian skipjack tuna fishery 1964-82. Administrative report H-86-14. Southwest Fisheries Center. U.S. Department of Commerce. Honolulu. 16 pp.

1987   Economic prospects for Hawaii's skipjack tuna industry. In Forces of change in Hawaii's aku industry 1986: A summary of a workshop. Edited by Christofer H. Boggs and Samuel G. Pooley. NOAA technical memorandum. National Marine Fisheries Service. Honolulu. In press.

Pooley, Samuel G.

1987   Economic profile of Hawaii's aku fleet. In Forces of change in Hawaii's aku industry 1986: A summary of a workshop. Edited by Christofer H. Boggs and Samuel G. Pooley. NOAA technical memorandum. National Marine Fisheries Service. Honolulu. In press.

State of Hawaii

1979   Hawaii fisheries development plan 1979. Division of Aquatic Resources. Department of Land and Natural Resources. Honolulu. 297 pp.

1985   State of Hawaii data book. Department of Planning and Economic Development. Honolulu. 662 pp.

1986   Hawaii fisheries plan 1985. Division of Aquatic Resources. Department of Land and Natural Resources. Honolulu. 163 pp.

Skillman, Robert A.

1987   Trends in Hawaii's aku production. In Forces of change in Hawaii's aku industry 1986: A summary of a workshop. Edited by Christofer H. Boggs and Samuel G. Pooley. NOAA technical memorandum. National Marine Fisheries Service. Honolulu. In press.

PART IV. REGIONAL AND INTERNATIONAL ASPECTS

14.
History and Role
of the Forum Fisheries Agency

Florian Gubon

INTRODUCTION

This paper reviews developments culminating in the formation of the South Pacific Forum Fisheries Agency (FFA).[1] It also examines the agency's role in the development and management of its members' fisheries resources and addresses the controversial question of whether FFA has a role in the management and conservation of highly migratory species of fish.

The paper has four parts. The first part discusses the concerns of South Pacific Forum (SPF) countries that prompted the establishment of a regional fisheries organization, showing that their action was largely a response to the deliberations of the Third United Nations Conference on the Law of the Sea (UNCLOS III).[2]

The second part examines FFA's role in fisheries management and development in the region by analyzing certain provisions of the FFA Convention.[3] This analysis is complemented by a review of the agency's work program. The third part addresses the specific question, Does FFA have a role in the management and conservation of highly migratory species of fish within the region? It argues that FFA does have such a role—one that is consistent with international law. The final section presents conclusions about the FFA.

ORIGINS

The idea of establishing a South Pacific fisheries agency originated at the seventh SPF meeting in Nauru in July 1976. It emanated from discussion of papers presented by Papua New Guinea and Fiji. Papua New Guinea's paper dealt with environmental conservation in the South Pacific, linking the fisheries question to the broader conservation issue. While acknowledging the work of the South Pacific Commission (SPC), the paper emphasized the roles of SPF countries as sovereign nations and the need for regional cooperation and coordination of activities related to the marine environment and its resources.

Fiji's paper, "The Law of the Sea," echoed some of the points of the Papua New Guinea paper, but it highlighted issues arising from the UNCLOS III negotiations. It pointed out that the concept of extended jurisdiction had been accepted internationally; it concluded by proposing the formation of a South Pacific regional fisheries organization. It suggested that another meeting be convened to examine the proposal in detail and specifically to discuss measures to enhance efforts in regional fisheries surveillance (South Pacific Forum 1976, 4–6).

The 1976 SPF meeting adopted a declaration—known as the Nauru Declaration—proposing that island governments take note of the importance of the UNCLOS III deliberations. In response to Fiji's initiative, it resolved to convene a meeting of officials to consider "the timing and terms of the creation of 200-mile zones; the problems and opportunities associated with them; the conservation of marine resources; the possible creation of a South Pacific fisheries agency; and the prospects for joint action and regional cooperation in matters such as surveillance and policing" (South Pacific Forum 1976, app. 4).

In October 1976 the officials met in Suva, where their governments (1) declared their intention to establish 200-mile exclusive economic zones (EEZs) after consulting with each other, (2) decided to harmonize fisheries policies in the region and to adopt a coordinated approach in their negotiations with distant-water fishing nations (DWFNs), and (3) decided in principle to establish a South Pacific fisheries agency to promote conservation and the rational use of tuna stocks in the region (Kearney 1978, 250). The spirit and intent of the Nauru Declaration were maintained, and it was endorsed by the Sixteenth South Pacific Conference and the SPF meeting in Port Moresby in August 1977.

At the Port Moresby SPF meeting, however, disagreement surfaced about the membership of the proposed fisheries organization. There were also conflicting opinions about the scope of cooperation among countries as delineated in the Informal Composite Negotiating Text (ICNT) of UNCLOS III. On the question of the organization's role, it was resolved that it should be limited to assisting member governments in exercising their sovereign fisheries management rights. Strong statements were presented that these rights should not be infringed in any way, even though it was recognized that conservation "was the concern of all countries in the region" (South Pacific Forum 1977, 27). Furthermore, SPF countries stressed that "the sovereignty of the coastal states should remain inviolable" (South Pacific Forum 1977, 37).

Concerning membership, the case was made that the proposed organization must not include DWFNs and that the United States in particular should not be permitted to participate (South Pacific Forum 1977, 31). The underlying problem was the U.S. tuna policy. SPF members saw this policy as being "directly opposed to that of the countries of the region and to the interests of the Pacific in particular" (South Pacific Forum 1977, 31).

Before resolving the membership issue, the countries agreed to convene an official meeting to (1) discuss and draft a convention establishing the proposed agency and (2) outline the agency's institutional arrangements.

The 1977 SPF meeting adopted the Port Moresby Declaration on the Law of the Sea and a Regional Fisheries Agency. The declaration proclaimed the intention of member governments of the forum to establish a South Pacific regional fisheries agency whose membership would be open to all forum countries and all countries in the South Pacific with coastal state interests who supported coastal state sovereignty to conserve and manage living resources—including highly migratory species—in the 200-mile zones (Kent 1980, 376).

In November 1977 a follow-up meeting took place in Fiji to initiate steps to implement the Port Moresby Declaration. Representatives of the United States, France, Japan, the United Kingdom, and Chile attended the meeting. Except for Japan, which had observer status, the nonforum countries attended purportedly to represent the nonsovereign territories they administered in the region. Because of the representation at this meeting—coastal states, metropolitan powers, and DWFNs—the nature of the proposed organization became a controversial issue. As Kent (1980, 376) has pointed out, two types of organizations were envisaged: one aiming primarily at ensuring conservation and promoting optimum use of living resources in the region, the other at ensuring maximum benefits for the peoples of the coastal states in the region and for the region as a whole.

Most SPF countries at the 1977 Fiji meeting expected that the proposed organization would be of the second type. However, as the meeting progressed, the United States was allowed to participate fully and spoke in support of its own interests as a DWFN and a major industrial fish-processing and fish-marketing nation rather than as a representative of the nonsovereign territories it administered. Largely because of the U.S. perspective, the meeting was unable to conclude a draft convention.

In June 1978 an officials' meeting was reconvened and a draft South Pacific Regional Fisheries Organization Convention was concluded. However, the compromises reached in the course of drafting the convention were unacceptable to the 1978 Niue SPF meeting. Concern was voiced that the draft convention proposed a "broadly based" Article

64-type of organization as envisaged in the ICNT. It was claimed that this type of organization was concerned primarily with fisheries conservation. The Niue SPF meeting noted that an Article 64 body is "fundamentally different from that originally envisaged by the Port Moresby Declaration" (South Pacific Forum 1978, 15). The Niue meeting was reminded that it had been agreed in Port Moresby to establish an organization that would enable island countries to band together to "present a united front in their negotiations with DWFNs" (South Pacific Forum 1978, 13-14).

The Niue meeting ended with a political directive that officials report on "the advisability and practicality of the draft convention together with any modification that may be desirable" (South Pacific Forum 1978, 26). A modified draft convention was then prepared and presented to the 1979 SPF meeting in Honiara. Drastically altered in intent, the draft convention was accepted by the political leaders.

The Nauru and Port Moresby declarations served as the basis upon which SPF countries established the FFA. By establishing the agency, the countries intended to manage the fisheries resources within their EEZs (including highly migratory species) so as to derive maximum financial benefits. An Article 64-type of body would not meet this goal.

FISHERIES MANAGEMENT AND DEVELOPMENT

The convention, signed in 1979, established FFA and serves as the legal instrument by which the agency exists and operates. In examining the FFA's role in fisheries management and development, the following provisions of the convention are reviewed: the Preamble—stating the objectives and concerns of parties to the convention; Articles IV and V—dealing with the establishment of the "Committee" and its functions; and Article VII—delineating the functions to be performed by the secretariat.

Concerns and Objectives

The Preamble to the FFA convention expresses the members' concerns and objectives: (1) their common interest in the conservation and optimum use of the living marine resources in the region, particularly highly migratory species; (2) their desire to promote regional cooperation and coordination in fisheries policies; (3) their concern for securing maximum benefits from the region's living marine resources for the national well-being of SPF countries; and (4) their desire to facilitate the collection, analysis, evaluation, and dissemination of information about the region's fisheries resources, especially highly migratory species.

The desire of SPF governments to adopt common measures both to conserve and to ensure optimum use of highly migratory resources

in the region stems largely from UNCLOS III. Under extended jurisdiction, coastal states were given sovereign rights to manage and conserve fisheries resources within their EEZs. The use of such expressions
as "conserve" and "optimum utilization" in the FFA convention raises
questions as to whether the agency is assuming the role of an Article
64-type of body. Van Dyke and Heftel (1981, 48–54) argue that the FFA
does not meet the Article 64 requirements, principally because of the
agency's membership. This issue is examined below; the point here is
that parties to the FFA convention were merely expressing "their common interests" about such matters when the convention was drawn
up and adopted.

The objective of SPF members in promoting regional cooperation
and coordination in fisheries policies is logical and understandable. In
1976 SPF members had been faced with claiming jurisdiction over large
tracts of the central and western Pacific Ocean—waters rich in fisheries
resources from which they could derive financial benefits. However,
Pacific island countries lacked the funds and the expertise needed to
exploit and develop these resources. Moreover, the vast size of the EEZs
of many countries presented surveillance problems. For these reasons
it was recognized that regional cooperation was imperative and that
it would yield benefits for all SPF countries.

After declaring their EEZs in conformity with prevailing international practice, SPF countries realized that they would have to negotiate fisheries access agreements with DWFNs because they lacked the
capacity to harvest their fisheries resources themselves. Furthermore,
they acknowledged that they lacked the experience and information
necessary to negotiate equitable agreements. Their size, geographic location, and level of economic development hindered them from tack-
ling these problems alone, but as a group they could achieve results
beneficial to all. Thus SPF members resolved that regional cooperation
was a feasible and sensible approach to solving their problems.

A major problem they faced in declaring extended jurisdictions and
entering into fisheries agreements with DWFNs was their limited information about the resources in their EEZs. Without this information
they were handicapped in making decisions about exploiting and developing their resources. They needed to know (1) the quantity and
quality of their fisheries resources; (2) how much fishing could be permitted while sustaining stocks; (3) who had traditionally fished their
resources; (4) where their fish was sold, processed, and consumed; and
(5) the market value of their fish. SPF members maintained that a
regional body such as the FFA would help them obtain some of this
information. This information would in turn enable them to exercise
management functions competently, both individually and as a group.

Forum Fisheries Committee

Article IV(1) of the FFA convention establishes a "Committee," which must meet at least annually. Article IV does not specify the composition of the committee, but it can be inferred from the words "each Party shall have one vote" (Paragraph 2) that each member is entitled to have at least one representative.

The Forum Fisheries Committee's functions are specified in Article V of the convention. These functions are to (1) provide detailed policy and administrative guidance and direction to the agency, (2) provide a forum for parties to consult on matters of common fisheries concerns, and (3) carry out other functions necessary to give effect to the convention (Paragraph 1).

Committee members are required to emphasize the promotion of regional cooperation in (1) harmonizing policies with respect to fisheries management, (2) maintaining relations with DWFNs, (3) surveillance and enforcement, (4) onshore fish processing, (5) fish marketing, and (6) cooperating with respect to fishing access to the EEZs of other FFA members (Paragraph 2).

Article V(1) prescribes the broad powers members have under the convention through the committee's role in directing and monitoring FFA activities. Paragraph 2 of the article specifies areas where members—through the agency—will promote coordination and cooperation in their activities.

The Forum Fisheries Committee has met annually since 1980. At its meetings it undertakes two primary functions: (1) it approves the director's report, which details the activities undertaken by the agency in the preceding financial year and the expenses incurred, and (2) it discusses and approves the agency's proposed work program for the next fiscal year. A typical work program contains information relating to (1) staffing, (2) harmonizing fisheries regimes and access agreements, (3) fisheries surveillance and enforcement, (4) information services, (5) tuna fishing development, (6) economic analyses, (7) fishing patterns, (8) fisheries training and administrative development, (9) the Regional Register of Fishing Vessels, (10) delineation of fishing and related zones, and (11) program management. If the committee accepts the director's report, it is presented several months later to the annual SPF meeting for approval. The committee is also entrusted with the task of appointing the director and deputy director on terms and conditions it considers appropriate.

Functions of the Agency

Article VII of the FFA convention sets forth the functions of the agency that are to be performed by the secretariat, subject to the direction of the committee. The agency's main tasks are to (1) collect, analyze, evaluate, and disseminate to members statistical and biological information concerning the region's living marine resources, particularly highly migratory species; (2) collect and disseminate to members information about management procedures, legislation, and agreements adopted by other countries, both within and beyond the region; (3) collect and disseminate to members information about fish prices and the shipping, processing, and marketing of fish and fish products; (4) provide to members, as requested, technical advice and information, assistance in developing fisheries policies and negotiations, issuing fishing licenses, collecting fees, or maintaining surveillance and enforcement; (5) seek to establish working arrangements with relevant regional and international organizations, in particular the SPC; and (6) undertake other functions determined by the committee.

FFA efforts to collect, analyze, and evaluate scientific and other fisheries information and to disseminate it to members have been admirable, given the agency's constraints. Although FFA has been unable to provide complete information, members have been able to obtain a clearer perception of the quantity and quality of fisheries resources within the region and to negotiate better financial deals in access negotiations with DWFNs.

More recently the FFA secretariat has concentrated on providing technical advice and assistance to members negotiating access agreements with DWFNs. In 1986, for example, FFA provided such assistance to all member countries. Negotiations it assisted in were Cook Islands/Korea; Federated States of Micronesia/Japan; Kiribati/Korea; Kiribati/Japan; Kiribati/USSR; Palau/Japan; Papua New Guinea/Japan; Tuvalu/Korea; Federated States of Micronesia, Kiribati, Palau/U.S. fishing interests; FFA states/U.S. government. To complement its assistance, FFA is putting effort into strengthening fisheries surveillance in the region to ensure that DWFN agreements are enforced and that illegal fishing is minimized.[4]

As part of its duties, FFA is required to establish working relationships with appropriate regional and international organizations. The agency has maintained good relationships with SPC, which has, through its Tuna and Billfish Assessment Program, provided FFA with basic information about the status of highly migratory fish stocks in the region. FFA also maintains close liaison with such United Nations agencies as the Food and Agriculture Organization (FAO) and the United Nations Development Program (UNDP). It also maintains links with donor agencies in several countries. However, it has taken no concrete measures to establish working arrangements with such international organizations as the Inter-American Tropical Tuna Commission, the

International Commission for the Conservation of Atlantic Tunas, and the Indo-Pacific Fisheries Council.

MANAGEMENT AND CONSERVATION OF HIGHLY MIGRATORY SPECIES

Accepted legal definitions of *management* and *conservation* with reference to highly migratory species do not exist. In the legal literature, the words have different meanings, but the difference is unclear. The Law of the Sea Convention uses both words collectively and individually but does not provide definitional assistance.

Since UNCLOS III, legal experts have attempted to define the meanings of *management* and *conservation* as used in the Convention (see, for example, Burke 1984 and Oda 1983). The UNCLOS III negotiators apparently assumed that the meanings of these critical terms were clear and therefore required no elaboration. Whether the lack of definitions was due to this assumption or to an oversight—deliberate or otherwise—on the part of the negotiators, the legal ambiguity of the terms makes it possible for lawyers and coastal states to argue cases beneficial to themselves.

The lack of definitional acceptance and clarity of *management* and *conservation* should not, however, hinder the exploitation and development of marine resources. A working understanding of the terms can be derived from their ordinary dictionary meanings—a detailed discussion of which is outside the scope of this paper.

Attempts to clarify narrow legal definitions and issues can lead to missing the woods for the trees. Legal ambiguities should be no excuse for not achieving the purpose of the law. The Law of the Sea Convention has clearly assigned coastal states the functions of both managing and conserving the fisheries—including highly migratory species—within their EEZs. Even Article 64 of the convention does not derogate this right: it requires coastal states to cooperate on management and conservation—which is what FFA member countries are doing.

A discussion of Articles III, VII, and IX of the FFA convention can clarify FFA's role with respect to highly migratory species of fish. Article III is perhaps the most important: Paragraph 1 mentions recognition by member countries of coastal state rights contained in Article 56 of the Law of the Sea Convention; Paragraph 2 records their recognition that for effective conservation and optimum utilization of the highly migratory species, an additional international body needs to be established to provide for cooperation between members and DWFNs.

Despite the importance of Paragraph 2, agency critics seem to deliberately ignore the purpose and intent of Article III. The paragraph states:

> Without prejudice to Paragraph (1) of this Article the Parties recognize that effective co-operation for the conservation and optimum utilization of the highly migratory species of the region will require the establishment of additional international machinery to provide for co-operation between all coastal states in the region and all states involved in the harvesting of such resources.

With respect to Article 64 obligations under the Law of the Sea Convention, Sutherland and Clark argue that FFA states have made, and continue to make, real attempts to meet their obligations to cooperate to establish such an organization (Sutherland 1985, 34–37; Clark 1986). They argue that the numerous unsuccessful regional meetings convened in attempts to establish a broadly based organization of the type envisaged by Article 64 are necessary and sufficient to meet the Law of the Sea Convention requirements.

Article VII sets out the functions of the agency, which include collecting, analyzing, evaluating, and disseminating to members information about highly migratory species of fish. This provision should be interpreted in the light of Articles III and IX. Article IX requires members to supply the agency with certain information, including catch-and-effort statistics on fishing operations in their waters and biological and statistical data. The agency analyzes and evaluates the information and disseminates it to members to enable them to exercise their management functions. The agency is clearly not vested with management decisions.

Lawyers generally agree that the implications of Article 64 of the Law of the Sea Convention are not readily understood, particularly when interpreted in the light of other EEZ provisions. Article 64 does not offer details on how "appropriate international organizations" are to be created, nor does it specify the functions of such bodies. In Burke's view, "it is up to the states directly concerned (coastal and fishing) to develop such international bodies as they can agree upon" (Burke 1984, 291). The FFA convention specifically provides for efforts to be made by members to establish an "appropriate international organization." The efforts have been made, but the establishment of such an organization ultimately depends on agreement among FFA members.

CONCLUSION

The formation of FFA was a deliberate attempt by SPF countries to capitalize on the rights and obligations granted them by the then-emerging Law of the Sea. They intended to create an organization unique to the region, one that would (1) provide a forum to discuss regional fisheries issues and (2) enable them to cooperate and coordinate their fisheries activities, with the ultimate goal of deriving maximum economic returns, both for themselves and for the region, from the resources within their EEZs.

FFA's essential role is coordination; it is not vested with making management decisions (for example, setting catch quotas). Such decisions are the responsibility of individual member countries, which are required to inform the agency about their management initiatives. It is primarily their exchange of ideas and information and their knowledge of what others are doing that provides the basis for the members' cooperative efforts. The agency's tasks are to help members cooperate and to give them technical advice and assistance upon request. This dual role makes the FFA a unique regional organization.

FFA carries out its functions through its work program. The agency has been effective in that real benefits have accrued to members as a result of its activities.

Although the FFA has achieved much, it faces financial constraints. The agency requires adequate funding to execute its programs and realize its goals. Two member countries—Australia and New Zealand—shoulder a large share of its financial burden. Together, they contribute two-thirds of FFA's annual budget, but there is no guarantee that they will do so indefinitely.

Other governments and organizations that support FFA's work in various ways include FAO, UNDP, the Japanese government, the Canadian International Development Agency, and the U.S. Agency for International Development. Help from these and other organizations has contributed to FFA's success. But again, there is no assurance, nor should FFA members expect that such assistance will always be forthcoming. The agency's members should therefore strive to operate within their own financial capabilities.

Financial assistance from countries and organizations outside the region brings with it the possibility that FFA's aspirations and objectives could be compromised. Assistance is often given to serve the donor's interests rather than the recipient's. Although this may not be the case with FFA donors, the possibility exists.

For Pacific island countries, the region's fisheries represent more than an economic asset. They are critical not only to national and regional politics but also to the culture of all Pacific islanders. "Fish is our life," island delegations told the U.S. delegation in Fiji in September 1984 when FFA countries first met with the U.S. government to try to negotiate a multilateral fisheries treaty.

Since time began, Pacific islanders have survived on islands and atolls by living primarily off what the ocean provides. That dependency remains; it has not diminished through time. Today, as young developing nations, Pacific island countries are turning to the oceans to try to strengthen their development prospects. It is no coincidence that these island countries should set the pace in regional fisheries cooperation. FFA has become a hallmark of cooperative efforts and success, and it provides a strong basis from which to face the challenges ahead.

NOTES

1. FFA's membership now includes Australia, Cook Islands, Federated States of Micronesia, Fiji, Kiribati, Nauru, New Zealand, Niue, Papua New Guinea, Solomon Islands, Tonga, Tuvalu, Vanuatu, and Western Samoa. The Marshall Islands and Palau have observer status, but for most practical purposes they are treated as full members.

2. The South Pacific Forum (SPF) was formed in 1971. Its members are Australia, New Zealand, and all independent and self-governing island countries in the region. It is a subgroup of the South Pacific Commission (SPC) formed to enable members to voice their joint political views denied to them by the SPC.

3. The South Pacific Forum Fisheries Agency Convention was concluded and signed by parties in July 1979 during the tenth SPF meeting in Honiara, Solomon Islands.

4. FFA's surveillance and enforcement project is financially assisted by the International Center for Ocean Development, Canada. The project, launched in November 1984, aims to (1) organize meetings of fisheries surveillance officials in the region and (2) conduct training courses and seminars for those working in fisheries surveillance.

REFERENCES

Burke, William T.

19821   U.S. fishery management and the new law of the sea. American Journal of International Law 76:24–55.

1984   Highly migratory species in the new law of the sea. Ocean Development and International Law 14(3):273–314.

Clark, Les

1986   Tuna industry development in the southwest Pacific. In Proceedings of the INFOFISH Tuna Trade Conference. Bangkok. pp. 151–159.

Kearney, R. E.

1978   The law of the sea and regional fisheries policy. Ocean
Development and International Law 5:249–286.

Kent, George

1980   Fisheries politics in the South Pacific. Ocean Yearbook
2:346–383.

Oda, S.

1983   Fisheries under the United Nations Convention on the
Law of the Sea. American Journal of International Law
77(4):739–755.

South Pacific Forum

1976   Summary of proceedings of the Seventh Meeting of the
South Pacific Forum. Nauru.

1977   Report of the Eighth Meeting of the South Pacific Forum.
Port Moresby, Papua New Guinea.

1978   Report of the Ninth Meeting of the South Pacific Forum.
Niue.

1979   Report of the Tenth Meeting of the South Pacific Forum.
Honiara, Solomon Islands.

Sutherland, William

1985   Regional cooperation and fisheries management in the
South Pacific. LL.M. diss. University of London.

Van Dyke, Jon, and Susan Heftel

1981   Tuna management in the Pacific: An analysis of the South
Pacific Forum Fisheries Agency. University of Hawaii Law
Review 3(1):1–65.

15.
Fisheries Cooperation:
The Case of the Nauru Group

David J. Doulman

INTRODUCTION

Tuna resources are distributed widely throughout the Pacific islands region. But because of differences in species distribution throughout the region and seasonal and overall differences in the distribution of particular stocks, not all island countries and territories have an equal interest in their commercial exploitation. It is the distribution of tuna in the region that (1) influences the posture of island governments toward distant-water fishing and domestic tuna industries and (2) largely determines the level, type, and distribution of distant-water fishing activity and affects the nature and extent of domestic tuna development within the region.

Soon after the 1979 inception of the Forum Fisheries Agency (FFA), several member countries with the greatest potential stake in the region's tuna fishery formed a subregional alliance known as the Nauru Group.[1] The alliance was based principally on economic considerations: these countries recognized that by harmonizing their relations with distant-water fishing nations (DWFNs) and cooperating on all tuna matters, they could derive maximum financial returns from the exploitation of their tuna resources. These principles of cooperation are consistent with FFA goals stated in the preamble of the agency's convention (SPFFA Convention 1979).

This paper traces the evolution of the Nauru Group, reviews its effectiveness with respect to its founding objectives, and analyzes its impact within the FFA. The paper commences with an account of the Nauru Group's formation. A review of the Nauru Agreement and a discussion of its implementing arrangements follows. The third section discusses the reaction of DWFNs to the group's formation. The impact of the group within the FFA is analyzed in the fourth section. Finally, some conclusions are drawn about the overall effectiveness of the group.

BACKGROUND AND RATIONALE

The Nauru Group consists of seven countries: Federated States of Micronesia, Kiribati, the Marshall Islands, Nauru, Palau, Papua New Guinea, and Solomon Islands. Recognizing the economic importance of the tuna resources within their exclusive economic zones (EEZs) and frustrated by the time it took (1976–79) to establish the FFA, representatives from the Federated States of Micronesia and Papua New Guinea conferred in Kiribati in 1980 at the annual South Pacific Forum meeting about forming a subregional tuna grouping.[2] Interest in the possibility of such a grouping grew, and an informal meeting was held in Port Moresby in 1980 when island countries gathered for the South Pacific Conference (Anderson 1982, 17). This meeting laid the foundations for a round of formal negotiations among the seven FFA member countries for the establishment of the Nauru Group.

The first round of negotiations was held in Nauru in March 1981. The second round, in Port Moresby eight months later, led to the completion of the Nauru Agreement text. At the end of the second round, the text was initialed by ministers holding fisheries portfolios in the Marshall Islands and Papua New Guinea and by senior officials from the Federated States of Micronesia, Kiribati, and Palau.[3] Representatives from Solomon Islands participated but did not initial the text of the agreement. Nauru's delegation was unable to attend the Port Moresby meeting and likewise did not initial the text.

Because they were eager to activate the agreement and to implement the cooperation it embodied, group members quickly obtained the necessary statutory approvals of their respective governments for the initialed text. Recognizing that delay would be costly in financial returns for the tuna being harvested by DWFN fleets in their EEZs, the seven countries signed the Nauru Agreement in Nauru in February 1982, only three months after its text had been agreed.

Despite the reservations of some FFA members about forming the Nauru Group, the alliance is considered logical and sensible for several important reasons:

• The group's members have contiguous EEZs (Figure 1). This fact has significant implications for resource management and for fisheries surveillance and enforcement of DWFN agreements. Moreover, the combined EEZs of the member countries cover about 14 million square kilometers of the central and western Pacific Ocean—45 percent of the total zone area of all Pacific island countries and territories and 72 percent of the zone area of FFA member countries (Table 1).[4]

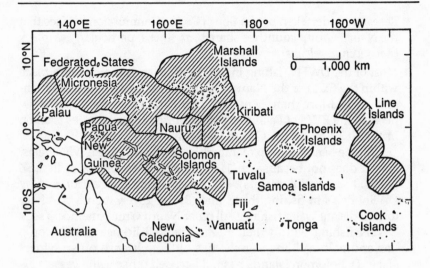

Figure 1. EEZs of the Nauru Group countries

Table 1. Nauru Group countries and their 1985 agreements with distant-water fishing nations.

Country	Zone size (million km²)	Fishing methods	Distant-water fishing nations
Federated States of Micronesia	3.0	a, b, c	d, e, f, g
Kiribati	3.6	a, b, c	d, f, h
Marshall Islands	2.1	a, b, c	d
Nauru	0.3	nil	nil
Palau	0.6	a, b, c	d
Papua New Guinea	3.1	a, b, c	d, e
Solomon Islands	1.3	a, b	d
Total	14.0		
Percentage of regional total	45		
Percentage of Forum total	72		

Source: Doulman 1986.

[a]longline. [b]pole-and-line. [c]purse seine. [d]Japan. [e]Taiwan. [f]Korea. [g]Mexico. [h]USSR.

- Research undertaken by the South Pacific Commission implies that the Nauru Group countries share tuna stocks, particularly skipjack (*Katsuwonus pelamis*).
- Most of the DWFN fishing in the Pacific islands region takes place within the EEZs of the Nauru Group countries and adjacent high-seas areas. More than 75 percent of the tuna harvested by DWFN fleets in the EEZs of Pacific island countries and territories is taken in the EEZs of Nauru Group countries.

All the Nauru Group members but Nauru license DWFN fleets. Nauru does not because of (1) the relatively small size of its EEZ (300,000 square kilometers), (2) the apparent lack of interest by DWFN fleets in fishing there, and (3) the government's passivity toward distant-water fishing.[5] All other Nauru Group members seek DWFN fishing within their EEZs. They all license DWFN long-line and pole-and-line vessels; five of them license purse seiners (Table 1). Solomon Islands has not licensed purse seine vessels except for a group seiner belonging to its joint-venture partner, the Taiyo Fishing Company of Japan, in the domestic tuna industry. However, following the 1986 conclusion of the regional tuna treaty for the U.S. purse seine fleet, Solomon Islands will be licensing U.S. seiners.[6]

In 1985 five DWFNs had fishing agreements with Nauru Group countries (Table 1). Except for Nauru, all member countries had access agreements with Japan; two countries (Federated States of Micronesia and Papua New Guinea) had agreements with Taiwan; two countries (Federated States of Micronesia and Kiribati) had agreements with Korea; Federated States of Micronesia had an access agreement with Mexico; Kiribati had one with the Soviet Union.

Other DWFN vessels also operated legally in the EEZs of Nauru Group countries on an ad hoc basis. For example, Papua New Guinea licensed a range of foreign fishing vessels under the terms and conditions of its Japanese agreement, in effect treating them as though they were of Japanese origin. The largest number of vessels licensed in this category were U.S. purse seiners.[7]

- The region's purse seine fleets—which in 1984 harvested an estimated 375,000 tonnes or 36 percent of the Pacific islands' total tuna catch—confine their operations to the Nauru Group waters, primarily to the zones of the Federated States of Micronesia and Papua New Guinea (Doulman 1986, 14). Depending on prevailing fishing conditions (which are a function of a variety of complex factors), up to 75 percent of annual reported distant-water purse seine catches are taken in the EEZs of these two countries.

- The bulk of Japan's distant-water tuna fleet—which accounted for 75 percent of total distant-water vessels deployed in the islands region in 1985—targets on the EEZs of Nauru Group countries and nearby high-seas areas (Doulman 1986, 8).

- Over 70 percent of DWFN access fee payments made under agreements in the Pacific islands go to Nauru Group countries. Total payments (excluding fisheries-related aid) to the region range between $20 and $25 million per year.[8] For some countries, DWFN access fee payments are a sizable proportion of government revenue. For example, about 25 percent of Kiribati's 1986 budget came from access fee payments (Doulman 1987). With few options for raising domestic revenue, Nauru Group countries must cooperate to protect this most valuable resource and to seek a fair financial return from its exploitation by DWFN fleets.

THE AGREEMENT

The Nauru Agreement is an international treaty that obligates members to adopt common courses of action with respect to their shared fisheries resources so long as the cooperation benefits them without derogating their sovereign rights.

Following the recitals, the agreement is divided into 11 articles (Appendix 1). The agreement broadly defines the areas of fisheries cooperation and harmonization that members should pursue, but the conclusion of subordinate arrangements (commonly known as implementing arrangements) is required to give the agreement effect.

Article I of the agreement briefly specifies its spirit and intent. Articles II and III outline areas of cooperation (for example, the determination of minimum terms and conditions of access for DWFN fleets to the EEZs of Nauru Group members and the standardization of licensing procedures). Articles IV and V focus on the group's relationship with the FFA and associated administrative arrangements.

Fisheries surveillance and enforcement issues are the subjects of Articles VI and VII. Articles VIII through XI cover a range of legal and technical issues, among them the relationship between the Nauru Agreement and other regional agreements and the conclusion of implementing arrangements.

Implementing Arrangements

One implementing arrangement has been concluded under the agreement. It specifies the minimum terms and conditions of fisheries access that member countries must adopt in licensing DWFN fleets.[9] These terms and conditions must be incorporated into all DWFN access agreements that Nauru Group members conclude.

These minimum terms and conditions are designed primarily to prevent DWFNs from trying to play one group member off against another when negotiating access rights—a practice that members agreed would erode their benefits from the exploitation of their tuna resources. Terms and conditions also simplify access negotiations. Specifying these terms and conditions removes them from the negotiating table, leaving fewer variables subject to negotiation and freeing the countries to devote their attention to more important issues, especially financial ones.

The Nauru Group's first implementing arrangement was concluded in 1982. Its five articles cover two important areas: (1) the Regional Register of Fishing Vessels and (2) licensing terms and conditions for DWFN fleets (licensing procedures, the rights of authorized personnel, requirements for reporting catch and maintaining fishing logs, and procedures for entering and exiting EEZs and for identifying licensed vessels).

All vessels licensed by Nauru Group countries must be listed on FFA's register before they are issued licenses under an access agreement. The primary purpose of the register is to maintain a check on DWFN vessels operating in the islands region. If a vessel infringes its terms and conditions of access in a Nauru Group country (or another FFA member country)—for example, by fishing in territorial waters or in a closed area of an EEZ—penalties can be imposed on that vessel. In the extreme, a vessel can be removed from the register, and no FFA member is permitted to license that vessel, even if it is sold, until its standing on the register is restored. Restoration of a vessel's standing could possibly be achieved by the owner's paying a fine and agreeing to abide by the terms and conditions of access thereafter.

The Regional Register and its sanctions have proven effective in enforcing compliance with the terms and conditions of DWFNs' access agreements with Nauru Group countries. In this way the register has helped to reduce the costs of administering agreements. The register has also reduced the incidence of illegal fishing and facilitated the collection of fines from DWFNs to prevent the deregistration of their vessels for violations. The most notable cases were in the Federated States of Micronesia with the U.S. seiners *Ocean Pearl* (1985) and *Priscilla M* (1986).

The Regional Register has been the most controversial aspect of the first implementing arrangement. Initially, the Japanese government and industry and the American Tunaboat Association (ATA) claimed that the register was illegal, discriminatory, and unenforceable. Japanese industry convened several meetings in late 1982 to discuss the register. It maintained that one country's imposing a penalty on a vessel for violating licensing conditions in another country constituted "illegal punishment" (Suisan Keizai 1982). Likewise, the ATA in its 1982 negotiations with Papua New Guinea and in its later negotiations with the Federated States of Micronesia, Kiribati, and Palau refused to acknowledge the register's existence or to take it seriously. By 1983, however, DWFNs were reluctantly beginning to accept the Regional Register as a necessity if they wanted to maintain existing arrangements or conclude new ones with Nauru Group countries.

Even though the minimum terms and conditions of the Nauru Agreement's first implementing arrangement are more lenient than conditions imposed on foreign fishing vessels in the EEZs of DWFNs, the DWFNs consistently argued that they were excessive, restrictive, and unreasonable. However, with time, DWFNs have accepted the terms and conditions as part of a standard access agreement package.

As the need arises, Nauru Group countries will conclude other implementing arrangements. These arrangements are likely to address (1) preferential access for fishing vessels from Nauru Group countries and (2) the standardization of methods for determining access fee payments for DWFN fleets.

Preferential access for fishing vessels from Nauru Group countries and access priority for particular categories of DWFN vessels will become more important as member countries develop and expand their longline and purse seine fleets or enter into special arrangements with DWFN vessel owners (for example, for basing vessels in member countries). Nauru Group countries have tentatively agreed that they will provide preferential access for their fleets to each other's EEZs. Some members are already doing this informally, but the practice remains to be formalized. DWFN vessels that base in Nauru Group country ports may be expected to request and to receive priority access to other members' EEZs.

The standardization of methods for determining access payments levied on DWFN fleets was achieved to some extent by 1986. All countries but Palau use a per-trip formula for calculating access fee payments (Doulman 1986, 22). Previously, a combination of per-trip payments and annual lump-sum payments was used. Palau is expected to adopt the per-trip system in 1987, thereby standardizing the methodology for determining access fees for all Nauru Group countries.

The per-trip method of licensing vessels was first used in Papua New Guinea and Solomon Islands in 1979 (Doulman 1986, 22). In 1984 it was extended to Federated States of Micronesia, Kiribati, and the Marshall Islands.[10] Initially, some Nauru Group members had reservations about the per-trip method of assessing access fee payments, believing it would increase administration costs and reduce their net returns. However, this was demonstrated not to be the case, and the per-trip system was popularly introduced.[11]

Further standardization in setting DWFN access fee payments—for example, by scaling rates according to the nature or scope of a DWFN's operation—is likely to be introduced in time. However, the variety of operational characteristics of DWFN fleets in the EEZs of Nauru Group countries is so great as to preclude precise standardization.

The needs of Nauru Group countries will determine the timing of the introduction and conclusion of subsequent implementing arrrangements. The countries have already agreed on the issues to be contained in the next two implementing arrangements, but no agreement has been reached as to when they will take effect.

The first implementing arrangement has demonstrated its value in simplifying access negotiations and protecting the interests of the group's members. Subsequent arrangements will be designed to achieve the same ends.

DWFN REACTION

DWFNs reacted apprehensively to the formation of the Nauru Group, coming as it did so soon after the formation of the FFA, which had also caused a degree of uneasiness. DWFNs believed the group would make uneconomic and unreasonable demands on their industries that would disrupt established fishing patterns and lead to conflict. DWFN fleets had experienced financial hardship from oil price increases in the 1970s. These increases, which structurally modified the fishing characteristics of all DWFN fleets, stemmed from the formation of an oil producers' cartel. In the face of a soft and weakening market for tuna—especially tuna for canning—DWFNs believed that the Nauru Group, by using a cartel approach, would attempt to extract excessive fees that fishermen could not reasonably bear.

Papua New Guinea was the first Nauru Group member to discuss the Nauru Agreement with a DWFN. This was done in Port Moresby in March 1982 with the Japanese (Matsuda and Ouchi 1984, 191). At this consultation, Papua New Guinea told the Japanese that discussion of the Nauru Agreement and its implications for future relations with DWFNs was a major point of the meeting (Papua New Guinea 1982, 15–20).[12] The Japanese responded that the agreement would have to be studied carefully by government and industry, warning that an excessive increase in access fees could cause Japanese fishermen to lose interest in fishing in the EEZs of Nauru Group countries.

At the same consultation, Papua New Guinea informed the Japanese that the Nauru Group countries were interested in exploring the introduction of a regional licensing system for all DWFN fleets operating in the EEZs of Nauru Group countries.[13] Papua New Guinea indicated to the Japanese that it favored implementing the proposed system gradually. Moreover, because fewer vessels were involved in the purse seine fishery than in the longline and pole-and-line fisheries, the purse seine fishery should be the first to change.[14] There was a precedent for this in that the U.S. fleet had been operating under a regional licensing system in the EEZs of three Nauru Group countries (Federated States of Micronesia, the Marshall Islands, and Palau) since 1980. Papua New Guinea believed that this system could be extended to cover all Nauru Group countries and all DWFNs. It was envisaged that the system would (1) be administered from a central office, (2) offer administrative advantages for both Nauru Group members and DWFNs, (3) provide more adequate compensation to island countries for the tuna harvested in their zones, and (4) yield better fisheries data for management (Papua New Guinea 1982, 17).[15]

Although DWFNs, particularly Japan, were understandably perturbed about the increasing cooperation among Pacific island countries, especially the concrete steps taken by the Nauru Group, their assessment of the overall impact of this cooperation was essentially incorrect. In analyzing the situation, DWFNs failed to appreciate not only the financial importance of access agreements to Nauru Group members but also their interest in fostering enduring and stable access arrangements. Group members were acutely aware that by cooperating they could pluck the DWFN goose but that they must not kill it in the process. Disruptions to fisheries access would be disadvantageous both to DWFN fleets and to the island countries licensing them. The formation of the Nauru Group and the implementation of its harmonization policies did not lead to the termination of any DWFN access agreement.

A major benefit accruing to members from the formation of the Nauru Group was that DWFNs could no longer play one island country off against another in negotiations. DWFNs had been able to negotiate favorable terms and conditions of access by claiming that (1) particular issues were inconsistent with DWFN agreements in force in neighboring countries, (2) they would break off negotiations and go to a neighboring country where they could obtain a better arrangement, and (3) they could not provide copies of their agreements with other island countries because the agreements were confidential. These tactics divided island countries, weakened their bargaining positions, and tended to create suspicions among them.

In the short term this negotiating strategy was effective, but it failed in the long term because it increased the perception by Nauru Group countries that they needed to collaborate closely on fisheries matters. Largely as a result of the group's formation, FFA representatives started to participate as advisers in renegotiating access agreements. DWFNs at first opposed this participation, but in time they saw it as inevitable and essential to maintaining access agreements. FFA participation in negotiations, the practice of Nauru Group countries exchanging details of negotiations and other information, and the exchange of access agreements themselves broke the "divide-and-rule" strategy of the DWFNs.

REGIONAL IMPACT

Before the Nauru Group formed, member countries of the FFA assessed the benefits and costs of joining the regional subgroup, realizing that splinter groups within the FFA could lead to fragmentation and dilute the agency's potency in regional and international fisheries affairs. However, the rationale for forming the Nauru Group was so overwhelming that the seven countries opted to form the alliance. Although forming the group was clearly in the members' self-interest, it also seemed that in the longer run the group could benefit the FFA membership as a whole.

Some countries—Papua New Guinea, for example—assessed the decision to join the Nauru Group on the basis of its potential impact on regionwide fisheries cooperation and broader regional issues. Papua New Guinea was anxious that it not be party to a divisive subregional grouping because it valued maintaining close ties with all South Pacific Forum member countries and fostering harmonious, genuine, and enduring regional cooperation. Convinced that the Nauru Group would not have a negative effect on regional cooperation, Papua New Guinea elected to join the group and even canvassed support for it from neighboring countries.

The Nauru Group meets at least once a year (usually prior to the annual meeting of the Forum Fisheries Committee) and at other times as members think necessary. Group meetings are characterized by frank, open, and wide-ranging discussions. Primarily for these reasons, the group has become an effective and progressive alliance.

In many respects the Nauru Group has become a trendsetter among FFA countries in determining relations between DWFNs and other FFA member countries. Cooperative policies adopted by the group are usually extended to all FFA countries fairly quickly. The Regional Register of Fishing Vessels is an example of an instrument developed by the Nauru Group and then extended to all FFA member countries.

Although the progressive policies of the Nauru Group have generally brought tangible benefits to FFA members, some FFA members have regarded it with a degree of uncertainty. Their uncertainty stemmed from (1) a belief that the Nauru Group could impede regional fisheries cooperation and erode regional solidarity and (2) a concern that FFA resources would support the activities of the group at the expense of the FFA membership as a whole. This uncertainty was held by countries that for resource or policy reasons had little direct interest in the activities of DWFN fleets in the region. Sensitive to this uneasiness, the Nauru Group has adopted a lower profile in recent years.

Contrary to popular belief, the Nauru Group is not an exclusive club: members may leave and new members may join (Articles X and XI of the Nauru Agreement). However, the distribution of tuna resources in the region (the factor motivating the formation of the group) is such that only two other Pacific island countries—Tuvalu and Vanuatu—are likely to consider joining the alliance.

As an alternative to the Nauru Group, countries in the southern and easterly part of the region could establish a regional subgroup of their own. An alliance based on the distribution of tuna resources in that part of the region and focused on the DWFN longline fishery could become a viable subgroup. The Nauru Group could serve as a model for establishing such an alliance.

Compelling resource and economic reasons stimulated the formation of the Nauru Group, but its composition is also interesting politically. The Nauru Group consists of five Micronesian countries (Federated States of Micronesia, Kiribati, the Marshall Islands, Nauru, and Palau) and two Melanesian countries (Papua New Guinea and Solomon Islands) that did not have a history of close cooperation. However, as a result of the alliance, cooperative efforts have been strengthened, not only in fisheries but in other areas as well. The thread that drew these countries together was their shared tuna resources. The formation of the Nauru Group has bound its seven members together in a way that probably would not have been achieved otherwise.

CONCLUSION

To obtain a fairer share of the benefits from the exploitation of their tuna resources by DWFN fleets, seven countries joined forces to establish the Nauru Group in 1982. The members of the group knew that it would be difficult for them to obtain increased benefits individually and that they needed to implement concrete cooperative measures under a single umbrella.[16]

The primary goal of the Nauru Group is to strengthen regional fisheries cooperation—a goal consistent with the principles upon which the FFA was established in 1979. The Nauru Agreement obligates group members to adopt common courses of action for their common good. Measures taken by the group have increased benefits to members; subsequent regional adoption of these policies has benefited all FFA members that license DWFN fishing fleets.

Initial DWFN suspicions about the FFA and the Nauru Group have now largely been allayed. In dealing with DWFNs, Nauru Group countries—principally out of self-interest—have not made unreasonable demands in pursuit of a larger return from the exploitation of their tuna resources. DWFNs are comfortable in their relations with Nauru Group countries, recognizing that the group's cooperative arrangements have not disrupted established fishing patterns.

As the need arises and as they see fit, Nauru Group countries will introduce new measures to enhance and extend fisheries harmonization. Like previously introduced cooperative measures, future measures should not alarm DWFNs. On the contrary, further standardization of terms of access for DWFN fleets should simplify the negotiation and administration of agreements to the advantage of both DWFNs and Nauru Group countries.

The close and genuine cooperation established by the Nauru Group exemplifies the degree of cooperation possible between Pacific island countries in pursuit of their common good. The effective functioning of the group within the FFA also demonstrates the flexibility of the agency's membership in accommodating the special interests and needs of its members.

NOTES

1. The Nauru Group is so named because the initial discussions concerning its formation took place in Nauru in 1980.

2. Pacific island countries had to contend with a range of difficult issues in establishing the FFA. Perhaps the most difficult issue was agency membership, particularly membership by the United States. For an analysis of this issue see Doulman 1984.

3. The title of the agreement is Nauru Agreement Concerning Cooperation in the Management of Fisheries of Common Interest.

4. The South Pacific Forum consists of independent Pacific island nations and some self-governing countries.

5. Nauru has an established interest in tuna fishing because it was the first country among Nauru Group members, indeed the first in the Pacific islands, to have its own purse seiners.

6. Solomon Islands has been reluctant to license purse seine vessels because of fears that their operations might adversely affect the country's domestic pole-and-line tuna industry. Solomon Islands' domestic tuna industry is nationally important, accounting for 41 percent of the country's exports in 1983 (Doulman and Kearney 1987).

7. The American Tunaboat Association (ATA), which has traditionally negotiated access agreements with coastal states in Latin America and the Pacific islands on behalf of the U.S. purse seine fleet, has not had an agreement in the islands region since 1985. Papua New Guinea has not had an agreement with the ATA since 1982.

8. Total annual payments vary depending on several factors, including movements in fish prices and the total number of fishing trips made per year.

9. The title of the arrangement is An Arrangement Implementing the Nauru Agreement Setting Forth Minimum Terms and Conditions of Access to the Fisheries Zones of the Parties.

10. As a matter of policy the Japanese are eager to change to a per-trip system of licensing throughout the Pacific islands region and have requested countries and territories that currently use the lump-sum system to change. The Japanese government in 1986 requested the Pacific French territories to consider adopting an agreement identical in nature and scope to the Papua New Guinea/Japan access arrangement.

    As early as 1982, Japanese industry informed Papua New Guinea that it wanted to change all its access agreements in the Pacific islands to the per-trip system for determining fees because (1) it provided flexibility for fishermen in determining where to fish, (2) fishermen knew the fees before they began each fishing trip, (3) fishermen regarded the system as fairer than the lump-sum system, and (4) it was easier for fishermen to pay on a per-trip basis (Papua New Guinea 1982, 15–16).

11. In contrast to the Japanese, the ATA refused to consider a per-trip system of licensing vessels in its agreements with Papua New Guinea (1982) and the Federated States of Micronesia, Kiribati, the Marshall Islands, and Palau (1983–84). The ATA maintained that this system would be confusing to U.S. vessel owners and therefore not acceptable (Doulman 1986, 27).

12. The fisheries access arrangement between Papua New Guinea and Japanese industry has been in force since 1981. Consultations under the arrangement are permitted when any party desires to clarify or discuss matters of mutual interest or concern. The consultations, although held in a formal atmosphere, are not negotiations per se. Rather, they are designed to resolve outstanding differences and to head off the need for formal renegotiation of the arrangement or, in the extreme, its termination. Through the consultative process, parties to the fisheries arrangement have been able to resolve differences and misunderstandings; as a result there has been a reasonable degree of cooperation between Papua New Guinea and Japan.

13. Although Papua New Guinea genuinely favored this licensing approach, it was obligated to raise the issue with the Japanese under the terms by which the United States had lifted Papua New Guinea's tuna embargo. This sanction had been imposed in line with provisions of the U.S. Fisheries Management and Conservation Act (1976) following Papua New Guinea's apprehension of the U.S. seiner *Danica* for illegal fishing. To have the embargo lifted, Papua New Guinea agreed to (1) remove the cause that led to its imposition (that is, to conclude an access agreement for the U.S. fleet), (2) explore the possibility of introducing a regional licensing agreement for DWFN fleets, and (3) release the confiscated vessel to its former owners on nondiscriminatory terms.

14. In 1982 the Japanese had 120 pole-and-line vessels, 700 longliners, 32 single purse seiners, and 7 group seiners that were licensed by Japan's Fishery Agency to operate as far south as Papua New Guinea (Papua New Guinea 1982, 4).

15. Papua New Guinea requested a meeting of the Nauru Group in Solomon Islands in 1982 to discuss the regional licensing concept. Papua New Guinea presented a paper outlining the operation of such a system. While embracing the concept, some members of the group felt it necessary to defer implementation of a regional licensing approach. Nonetheless, the provisions of the 1986 U.S. tuna treaty with Pacific island countries closely resemble the proposal made in Papua New Guinea's 1982 paper.

16. Nauru, Papua New Guinea, and Solomon Islands formally declared their EEZs (fishing and/or economic zones) in 1978; Federated States of Micronesia, Marshall Islands, and Palau in 1979; and Kiribati in 1983 (Doulman 1986, 2).

REFERENCES

Anderson, David

1982   Developing island states move to protect tuna. Papua New Guinea Foreign Affairs Review 2(1):13–17.

Doulman, David J.

1984   The development of Papua New Guinea's domestic tuna fishery: A proposal for future management. Ph.D. diss. James Cook University. Australia. 536 pp.

1986   Fishing for tuna: The operation of distant-water fleets in the Pacific islands region. Research report series no. 3. Pacific Islands Development Program. East-West Center. Honolulu. 38 pp.

1987   The Kiribati-Soviet Union fishing agreement. Pacific Viewpoint. Wellington, New Zealand. In press.

Doulman, David J., and Robert E. Kearney

1987   The domestic tuna industry in the Pacific islands region. Research report no. 7. Pacific Islands Development Program. East-West Center. Honolulu. In press.

Matsuda, Y., and K. Ouchi

1984   Legal, political, and economic constraints on Japanese strategies for distant-water tuna and skipjack fisheries in Southeast Asian seas and the western central Pacific. In Memoirs of the Kagoshima University Research Center for the South Pacific 5(2):151–232. Kagoshima.

Nauru Agreement

1982   Nauru Agreement Concerning Co-operation in the Management of Fisheries of Common Interest. Forum Fisheries Agency. Honiara, Solomon Islands. 11 pp.

Papua New Guinea

1982   Record of discussions between the government of Papua New Guinea, the government of Japan, and the Japanese fishing associations, July 27–29. Division of Fisheries. Department of Primary Industry. Port Moresby. Mimeo. 28 pp.

South Pacific Forum Fisheries Agency

1979   South Pacific Forum Fisheries Agency Convention. Forum Fisheries Agency. Honiara, Solomon Islands. 6 pp.

Suisan Keizai
  1982   Japanese fishery organizations discuss Nauru Agreement.
              Tokyo. November 30. Translated from the Japanese by the
              Papua New Guinea Embassy, Tokyo.

APPENDIX

Appendix 1. The Nauru Agreement

NAURU AGREEMENT CONCERNING
CO-OPERATION IN THE MANAGEMENT
OF FISHERIES OF COMMON INTEREST

The Federated States of Micronesia, the Republic of Kiribati, the Marshall Islands, the Republic of Nauru, the Republic of Palau, Papua New Guinea and Solomon Islands:

TAKING into account the work of the Third United Nations Conference on the Law of the Sea;

NOTING that in accordance with the relevant principles of international law each of the Parties has established an exclusive economic zone or fisheries zone (hereinafter respectively called "the Fisheries Zones") which may extend 200 nautical miles from the baselines from which their respective territorial seas are measured and within which they respectively and separately exercise sovereign rights for the purposes of exploring, exploiting, conserving and managing all living marine resources;

HAVING REGARD to the objectives of the South Pacific Forum Fisheries Agency Convention and in particular the promotion of regional co-operation and co-ordination of fisheries policies and the need for the urgent implementation of these objectives through regional or sub-regional arrangements;

CONSCIOUS of the exploitation of the common stocks of fish, both within the Fisheries Zones and in the water adjacent thereto, by the distant water fishing nations;

MINDFUL of their dependence, as developing island states, upon the rational development and optimum utilisation of the living resources occurring within the Fisheries Zones and in particular, the common stocks of the fish therein;

RECOGNISING that only by co-operation in the management of the Fisheries Zones may their peoples be assured of receiving the maximum benefits from such resources; and

DESIROUS of establishing, without prejudice to the sovereign rights of each Party, arrangements by which this may be achieved;

HAVE AGREED AS FOLLOWS:

273

ARTICLE I

The Parties shall seek, without any derogation of their respective sovereign rights, to co-ordinate and harmonise the management of fisheries with regard to common stocks within the Fisheries Zones, for the benefit of their peoples.

ARTICLE II

The Parties shall seek to establish a co-ordinated approach to the fishing of the common stocks in the Fisheries Zones by foreign fishing vessels and in particular:

(a) shall establish principles for the granting of priority to applications by fishing vessels of the Parties to fish within the Fisheries Zones over other foreign fishing vessels;

(b) shall establish, as a minimum, uniform terms and conditions under which the Parties may licence foreign fishing vessels to fish within the Fisheries Zones regarding:

    (i) the requirement that each foreign fishing vessel apply for and possess a licence or permit;

    (ii) the placement of observers on foreign fishing vessels;

    (iii) the requirement that a standardized form of log book be maintained on a day-to-day basis which shall be produced at the direction of the competent authorities;

    (iv) the timely reporting to the competent authorities of required information concerning the entry, exit and other movement and activities of foreign fishing vessels within the Fisheries Zones; and

    (v) standardized identification of foreign fishing vessels;

(c) seek to establish other uniform terms and conditions under which the Parties may licence foreign fishing vessels to fish within the Fisheries Zones, including:

    (i) the payment of an access fee, which shall be calculated in accordance with principles established by the Parties;

    (ii) the requirement to supply to the competent authorities complete catch and effort data for each voyage;

    (iii) the requirement to supply to the competent authorities such additional information as the Parties may determine to be necessary;

    (iv) the requirement that the flag State or organisations having authority over a foreign fishing vessel take such measures as are necessary to ensure compliance by such vessel with the relevant fisheries law of the Parties; and

(v) such other terms and conditions as the Parties may from time to time consider necessary.

ARTICLE III

The Parties shall seek to standardize their respective licensing procedures and in particular:

(a) seek to establish and adopt uniform measures and procedures relating to the licensing of foreign fishing vessels, including application formats, licensing formats and other relevant documents; and

(b) explore the possibility of establishing, without prejudice to the respective sovereign rights of the Parties, a centralised licensing system of foreign fishing vessels.

ARTICLE IV

The Parties shall seek the assistance of the South Pacific Forum Fisheries Agency in establishing procedures and administrative arrangements for the exchange and analysis of:

(a) statistical data concerning catch and effort by fishing vessels in the Fisheries Zones relating to the common stocks of fish; and

(b) information relating to vessel specifications and fleet composition.

ARTICLE V

1. The Parties shall seek the assistance of the South Pacific Forum Fisheries Agency in providing secretariat services for implementing and co-ordinating the provisions of this Agreement.

2. An annual meeting of the Parties shall be convened preceding or following the regular session of the Forum Fisheries Committee in order to promote the implementation of this Agreement. Additional meetings may be convened at the request of three or more Parties. Such requests shall be communicated to the Director of the Forum Fisheries Agency who will inform the other Parties.

3. With the concurrence of the Parties, members of the South Pacific Forum Fisheries Agency, not Parties to this Agreement, may attend, as observers, the meetings referred to in this Article.

ARTICLE VI

The Parties shall, where appropriate, co-operate and co-ordinate the monitoring and surveillance of foreign fishing activities by:

(a) arranging for the rapid exchange of information collected through national surveillance activities;

(b) exploring the feasibility of joint surveillance; and

(c) developing other appropriate measures.

ARTICLE VII

The Parties shall seek to develop co-operative and co-ordinated procedures to facilitate the enforcement of their fisheries laws and shall in particular examine the various means by which a regime of reciprocal enforcement may be established.

ARTICLE VIII

Nothing contained in this Agreement shall be construed as a derogation of any of the rights and obligations undertaken by any of the Parties under the South Pacific Forum Fisheries Agency Convention or any other international agreement in effect on the date on which this Agreement enters into force.

ARTICLE IX

The Parties shall conclude arrangements where necessary to facilitate the implementation of the terms and to attain the objectives of this Agreement. The Parties concluding such arrangements shall lodge copies with the depositary of this Agreement.

ARTICLE X

1. This Agreement shall be open for signature by the States named in the preamble hereto and shall be subject to ratification.

2. This Agreement shall enter into force thirty days following receipt by the depositary of the fifth instrument of ratification. Thereafter it shall enter into force for any signing or acceding State thirty days after receipt by the depositary of an instrument of ratification or accession.

3. This Agreement shall be deposited with the Government of Solomon Islands which shall be responsible for its registration with the United Nations.

4. Following entry into force, this Agreement shall be open for accession by other States with the concurrence of all of the Parties to this Agreement.

5. Reservations to this Agreement shall not be permitted.

ARTICLE XI

1. This Agreement is a binding international agreement concluded among States and is governed by international law.

2. Any Party may withdraw from this Agreement by giving written notice to the depositary. Withdrawal shall take effect one year after receipt of such notice.

3. Any amendments to this Agreement proposed by a Party shall only be adopted by unanimous decision of the Parties.

   IN WITNESS WHEREOF the undersigned, duly authorised by their respective Governments, have signed the Agreement.

   DONE at Nauru this eleventh day of February One Thousand Nine Hundred and Eighty Two.

For the Government of the Federated States of Micronesia

For the Government of the Republic of Kiribati

For the Government of the Marshall Islands

For the Government of the Republic of Nauru

For the Government of the Republic of Palau

For the Government of Papua New Guinea

For the Government of Solomon Islands

16.
Global Tuna Markets:
A Pacific Island Perspective

Dennis M. King

INTRODUCTION

The Solomon-Taiyo joint venture in Solomon Islands prepares a shipment of canned tuna for delivery to a buyer in London; the product will be sold in the United Kingdom under the John West label. Te Mautari, Ltd., the Kiribati national fishing company, prepares a shipment of tuna for delivery to the Star-Kist cannery in American Samoa, where it will be canned, shipped to the United States, and sold under the Star-Kist label. Papua New Guinea collects access fees from a U.S. tuna purse seiner; its catch will be delivered to Thailand, where it will be canned and shipped to France to be sold under the Le Nonpareil label.

In each instance, a Pacific island nation is engaging in the international tuna trade, but the transactions take place at different market levels. The financial investment required, the risk involved, and the potential for economic payoff differ vastly across the levels. However, whether an island nation is selling access fees to a U.S. fisherman, raw or frozen tuna to a Thai canner, or processed-tuna products to a European wholesaler, the financial return and the economic impact of the transaction will be influenced by competitive forces in global tuna markets.

Consider, for instance, the impact in each instance if a new cannery in Thailand begins to penetrate U.S. and European markets by offering large volumes of low-priced canned tuna to buyers in those markets. To remain competitive in the United Kingdom, Solomon-Taiyo would have to reduce its canned-tuna price to its London buyer. In searching for alternative markets, Solomon-Taiyo would learn that major canned-tuna wholesalers around the world have already developed reliable sources of supply and will pay no more for Solomon Islands canned tuna than for comparable products from Thailand. Star-Kist, feeling similar competitive pressures, might drop its price in the U.S. canned-tuna market, but it is also likely to lower the price it offers to

279

Kiribati and other suppliers of raw or frozen tuna to offset the reduction in revenues. Te Mautari can look for a better price elsewhere, but it is likely to find that most canners have reliable sources of raw or frozen supplies and will buy tuna from new suppliers only if the price is lower than Star-Kist is paying. The downward pressure on raw or frozen tuna prices would also lower the expected earnings of the U.S. purse seiner operating in Papua New Guinea waters, thus reducing the amount it could afford to pay in access fees.

In each of these hypothetical situations, an island nation engaging in the global tuna trade is affected by a sequence of market changes beyond its control. Yet each nation must understand the structure of global tuna markets and how market forces are transmitted from one market level to another and from one region to another if it is to assess the financial risks involved in each segment of the tuna industry and the competitive pressures that influence the behavior of foreign business partners and competitors. Competition in the global tuna industry is fierce at all market levels, and no one can hope to succeed in the tuna business—as a landlord selling resource access, as a trader, or as an industrialist—without knowing how competitive pressures are transmitted between market levels. This paper describes the important characteristics of world tuna markets from the Pacific island perspective. Some market statistics are provided for illustration. However, because global tuna markets change constantly, the focus is on the trading system rather than on specific trading activities.

NEW OPPORTUNITIES

Along with internationally recognized jurisdiction over the commercially important tuna resources of the central and western Pacific come new economic opportunities for island nations. Tuna resources, however, have no commercial value until they are harvested, processed, and delivered to market. The critical question facing each Pacific island nation, therefore, is what role to play in the complex industry that takes tuna from the ocean, processes it, and delivers it to markets around the world.

Each island nation has both opportunities and limitations in the tuna industry. A nation's proximity to major tuna fishing areas is only one of the factors influencing its competitive position in the industry. Access to capital and to technical, business, and marketing skills is much more important than easy access to tuna resources. Yet it is in capital and in skills that most Pacific island nations remain disadvantaged. The real economic challenge for most of them, therefore, does not lie in exploiting their own tuna resources but in exercising control over their tuna fisheries so as to generate maximum economic benefits in ways consistent with their cultures and with regional economic development. This goal may not require that each island nation be directly involved in the tuna industry or in tuna markets.

Rather than investing their own scarce development funds in expensive and risky tuna operations, most island nations would probably be better off earning foreign exchange by licensing foreign operators, perhaps as part of a regional licensing scheme. Large-scale tuna operations based in remote island settings tend to attract people from outlying districts, frequently generating undesirable demographic shifts and cultural and social changes. Furthermore, because most materials for tuna operations—such as fuel, equipment, nets, and cans—are imported, locally based tuna operations do not usually yield significant gains in foreign exchange, nor do they result in economic self-sufficiency. In general, tuna production is a capital-intensive industry requiring an infrastructure or resource base that many island nations do not have. However, whether an island nation chooses to invest in fishing operations itself or chooses a less risky option, it must deal with foreign businesses that control large-scale harvesting and processing operations and that have access to major tuna markets. To deal effectively with them requires that island leaders be informed about tuna production systems, the international tuna trade, and ways of determining the value of tuna at different market levels.

TUNA PRODUCTION SYSTEMS

A global tuna production system can be thought of as a sequence of operations by which tuna is taken from the ocean, made into products, and delivered to consumer markets (Figure 1). These operations include harvesting, transshipping, processing, distributing, and marketing, each of which may be done independently by different operators. The restructuring of the global tuna industry since 1981 reflects geographical shifts in operations of some of the world's largest corporate production systems and some important changes in the way operators at different stages are linked. Market transactions, for instance, have replaced contractual or financial linkages as the chief method of passing tuna from one production stage to the next. Although there have been no major changes in the way each operation is performed, the need for efficiency and reliability is greater now because of increasing competition among firms operating at each stage. With fewer long-term contracts linking operations at each stage, the risks inherent in investing in any specific venture have also increased, especially in tuna-harvesting operations.

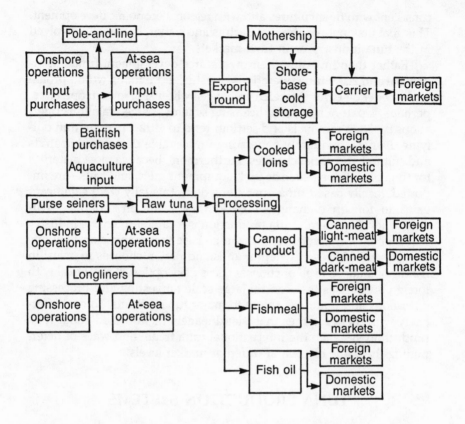

Figure 1. Pathways of commercial tuna development

Until recently most tuna fishermen around the world were affiliated through financial and equity-sharing arrangements with one of a few major processors who not only handled the distribution and marketing of tuna products but also coordinated the whole production system, from delivering raw fish to processors to advertising tuna products in markets. Boat owners and fishermen were responsible primarily for locating and capturing tuna; they were only peripherally involved in the complex logistical arrangements linking the various operations into a cohesive production system.

In the modern global system, however, there is much less integration of geographically diverse operations by the major processing companies. Large processors still attempt to regulate the flow of raw tuna into the production system and the distribution of processed tuna from it, but there is a new element of competition within each production system as well as between the corporations that coordinate the production systems. Independent operators at each stage (harvesting, processing, distribution, and so on) compete with each other and with operators at all other stages for a share of the industry's profits. Each dollar paid to a wholesaler for canned tuna is a dollar less for the retailer; each dollar paid to fishermen for raw or frozen tuna is a dollar less for the processor; each tonne of tuna sold by one fisherman is a tonne less that processors will need from other fishermen. The modern tuna market is competitive at many levels and, as Table 1 shows, the amount of profit to be passed around is limited.

Table 1 shows how tuna gains economic value as it passes through the production system from the primary producer (dollar value to the fisherman) to the final market (dollar value to the retailer). The dollar values in Table 1 are for illustration only, but because the value added at each stage barely exceeds the cost of passing tuna through that stage, the profits available to operators at each stage are small. Although the values shown are for illustration only, they represent conditions in the modern tuna industry that make it difficult for operators at any stage to keep costs low enough to compete successfully and earn a profit.

If their domestic tuna industries are to succeed, Pacific island nations must develop a "comparative advantage" in the production of raw or frozen tuna, tuna loins, or canned tuna. Unless they can achieve an advantage that makes them competitive with other tuna-producing nations, they will not be worthwhile partners in a tuna production-distribution-marketing system, nor will they succeed in attracting committed foreign investors. Island nations must determine what kinds and how many operations of a tuna production system they can perform and who should help them learn to perform these operations. Moreover, they need to decide which national goals they can achieve by participating in the global tuna industry: creating jobs, raising incomes, producing foreign exchange, developing technical skills, and the like. National leaders must recognize that competitive forces will prevail in the global tuna industry and that they may never realize the financial returns and economic effects of taking on one or more operations in a tuna production system.

TUNA MARKETS

In recent years the United States, Japan, and Europe have accounted for over 80 percent of global tuna consumption (Table 2). Tuna consumed outside these markets is most often produced by local or subsistence fisheries, which may be important to the region but have little impact on the world tuna trade (Figure 2).

Table 1. Value added in a tuna production system

Stage of production	Description	$ per tonne[a]			
		Market value[b]	Value added[c]	Production cost[d]	Potential profit[e]
Harvesting	Catch tuna, freeze, and deliver to nearby port	650	650	620	30
Transshipping	Collect tuna and deliver to processor.	775	125	110	15
Processing	Process (canned tuna, pet food, etc.).	1,150	375	300	75
Distribution	Deliver products to market	1,250	100	90	10
Marketing	Market and promote brand identification (advertising, promotional allowances, and wholesale discounts.	1,450	200	190	10
Retailing	Provide shelf space, promotions, and final market.	2,500	1,050	850	200
	Final market value equal to value added at each stage ($2,500). Total production cost ($2,160) is 86 percent of total value added. Potential profit ($340) is 14 percent of value added.	2,500	2,500	2,160	340

[a]Dollar figures are for illustration only.

[b]Market value is the approximate value after the stage of production.

[c]Value added is the difference in market value before and after the stage of production.

[d]Production costs are costs incurred in passing tuna through each stage of production.

[e]Potential profit is the amount earned by operators at each stage of production if the value added exceeds the costs at that stage. This potential profit is the basis of competition in the industry and provides the incentive for independent operators to invest at each stage.

Table 2. Major participants in the global tuna trade 1985

Producing nations				Importing nations				Consuming nations	
Major tuna harvesters	% of total	Major canned tuna producers	% of total	Raw/frozen tuna importers	% of total	Canned tuna importers	% of total	Canned tuna consumers	% of total
Japan	38	USA	36	Thailand	28	USA	46	USA	52
USA	13	Japan	17	USA	25	France	18	Japan	13
Spain	6	Thailand	13	Japan	24	UK	11	France	9
Indonesia	6	Italy	9	Italy	14	W. Germany	6	Italy	8
France	5	France	6	Ivory Coast	4	Belgium/ Luxembourg	2	Mexico	4
Taiwan	5	Spain	4	France	3	Italy	2	Spain	4
Philippines	5	Mexico	4	Spain	2	Other Europe	7	UK	4
Mexico	4	Ivory Coast	3	Other	1	Canada	5	W. Germany	2
Korea	3	Philippines	3			S. Africa	1	Other	7
Thailand	2	Indonesia	2			Australia	1		
Other	14	Taiwan	2			Other	1		
		Other	1						
Total[a]	100	Total	100	Total	100	Total	100	Total	100

Total tuna harvest 2.1 million tonnes

Total canned tuna production 78 million standard cases

Total tuna trade 850,000 tonnes

Total canned tuna trade 23.5 million standard cases

Total canned tuna consumption 78 million standard cases

[a] Totals may not add to 100 percent due to rounding.

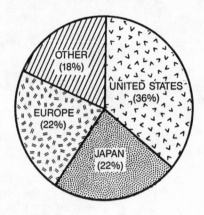

Figure 2. Global tuna market

The tuna market in Japan is much more intricate and volatile than markets in the United States and Europe, but the underlying forces of the global market prevail in all major national markets. It should be noted in Figure 3 that, although monthly fluctuations in tuna prices are more erratic in Japan than in the United States, the price trends (as reflected by a 24-month moving average) are almost identical.

Although the Japanese tuna market is two-thirds the size of the U.S. market and larger than the European market, it is highly specialized. Most tuna in Japan is eaten raw as sashimi; the remainder is sold primarily dried, smoked, pickled, or fermented. Although the celebrated Japanese markets for high-priced tuna are commercially important, over 80 percent of the global tuna harvest is destined for the canned-tuna trade, and the Japanese market has little impact on the market for cannery-grade tuna. In the United States and Europe virtually all tuna is sold in cans; for most purposes, it is the global market for canned tuna that is important to Pacific island nations (Figure 4). During 1985 the United States accounted for nearly half the world market for canned-tuna exports and an equal share of the export market for cannery-quality raw or frozen tuna. To appreciate the importance of the U.S. market on the world tuna trade, consider that a decline of a tenth of a kilogram in 1985 U.S. per capita tuna consumption (from 1.5 kilograms to 1.4 kilograms) would constitute an overall decline in U.S. canned-tuna sales of about 2.5 million cases (nearly $100 million

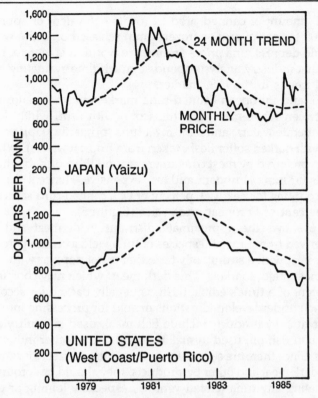

Figure 3. Trends in U.S. and Japanese skipjack tuna markets

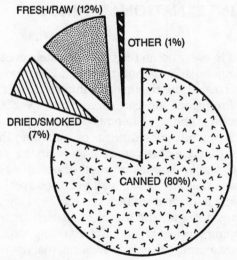

Figure 4. Final market tuna product forms

retail). This amount of canned product (or its equivalent of approximately 40,000 tonnes of raw or frozen tuna) released onto the world market would depress tuna prices around the world. It is therefore not surprising that industry analysts spend so much time evaluating conditions and trends in the U.S. market.

The U.S. and European canned-tuna markets make a major distinction between white-meat tuna (albacore, with firm, white flesh, taken in temperate waters) and light-meat tuna (primarily skipjack and yellowfin, with darker, softer flesh, taken from tropical waters). White-meat tuna is preferred by most consumers around the world, but light-meat tuna is far more abundant and less expensive. Hence it is more popular, accounting for nearly 80 percent of U.S. canned-tuna sales and nearly 90 percent of European tuna sales (Figure 5).

Figure 6 shows the approximate distribution of meat and by-products derived from tuna. All species of tuna yield a certain percentage of dark meat with a strong, oily taste that is not popular with consumers in most major markets. This dark meat, which may constitute over 20 percent of a tuna's edible flesh, is usually packed for secondary markets in underdeveloped nations or sold for processing into pet food. Other tuna by-products include fish meal, used in poultry and mink feed, and fish oil, used in making paint and in other industrial applications. Since there are secondary international markets for canned and frozen dark meat and other by-products, they should be accounted for in evaluating any tuna-related venture, especially loining or canning operations.

INTERNATIONAL TRADE

Supply Sources

Tuna are harvested in every ocean (Figure 7) and delivered to processing sites and consumer markets around the world (Figure 8). There are many species of tuna, but the five major commercial species (yellowfin, skipjack, bluefin, bigeye, albacore) account for over 90 percent of the international trade. Although different species are popular in different markets, they frequently substitute for one another; thus the harvest of one species from one ocean area competes in the world market with the harvest of similar species from other ocean areas.

Modern telecommunication systems allow seafood brokers, traders, and corporate buyers in diverse countries to receive daily information about prices, supply, demand, and inventory in each regional market. Fast, efficient shipping services link major harvesting, processing, and consumption centers around the world, ensuring that few international

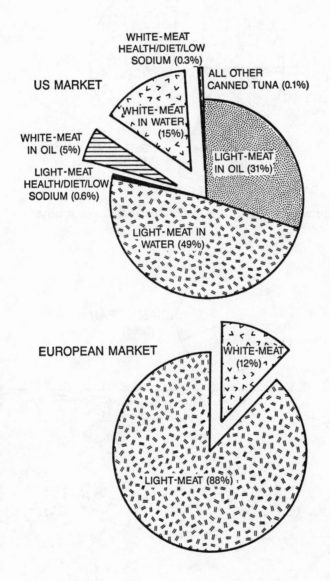

Figure 5. US /European canned tuna markets

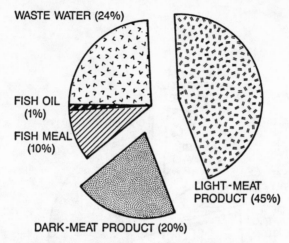

Figure 6. Approximate yield from yellowfin and skipjack tuna

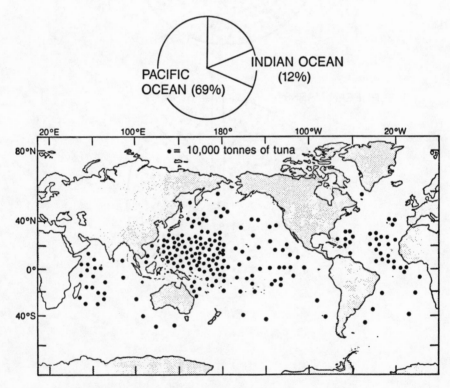

Figure 7. Tuna harvesting locations

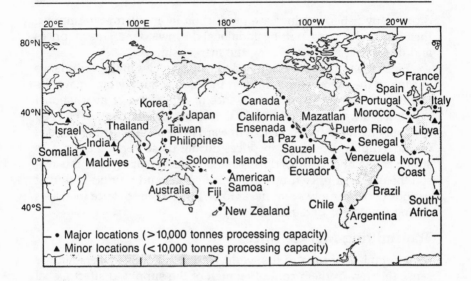

Figure 8. Tuna-processing locations

trading opportunities are overlooked. Although shipping costs affect international price competition, the important factors in determining the terms of trade are the supplier's reputation for quality and reliability, the buyer's integrity, and the effect of currency exchange rates and international tariffs.

Market Forces

Although market conditions change constantly, world demand for tuna, like the demand for other highly valued food products, is relatively stable and insensitive to price changes. Even if sales of a single brand respond to a price change, tuna sales industrywide are relatively inelastic. Brand loyalty among consumers is weak, so that a change in the market share of one brand is offset by a compensating change in the market share of a competing brand, causing little change in the size of the overall market. What is true for a particular brand, in other words, is not necessarily true for the industry and, despite the experience of individual firms, across-the-board increases in retail prices result in relatively small declines in the volume of sales; more important, industrywide price cuts result in relatively small increases in the amount of tuna sold. Thus when there is an oversupply of tuna on the world market, it is difficult for the industry to expand sales, and inventories build up quickly. Tuna prices can fall drastically as firms attempt to unload their inventories, and production cutbacks and idle capacity in the harvesting and processing sectors are inevitable. Ironically, as tuna prices drop, many fishermen increase production in an attempt to improve their financial returns, putting additional downward pressure on prices and deepening and prolonging the market slump.

Since the tuna supply is linked to natural cycles in the abundance of tuna in the ocean, large shifts in the market supply are unavoidable; hence tuna prices, in the absence of some kind of international cartel, will always tend to be cyclical. Because inventories can be used to supply markets and moderate upward price swings during periods of undersupply, upswings in prices during these periods are not as strong as the downswings during periods of oversupply. For this reason the effects of price drops felt by fishermen and producers during periods of oversupply seem never to be fully offset by price increases during periods of undersupply.

Transmitting Market Forces

Most suppliers of tuna to the world market are "price takers" in the sense that few of them control enough of the supply to affect global or even regional tuna prices. Although the global supply influences regional tuna prices, the regional supply has little effect on global—or even regional—prices. Poor fishing conditions in the western Pacific, for example, do not produce a significant increase in prices paid to Korean fishermen who operate in the western Pacific and sell their catch on the world market. The fact that prices are determined by global supply and demand means that export-oriented producers are in a risky position. Though it is possible that local catches will be high when international prices are high because of an undersupply on the world market, it is also possible that local catches will be low when prices are low because of an oversupply. Thus fishermen who produce for the global market are subject to much greater risk than fishermen who produce only for local markets, where prices respond to changes in supply. The global nature of tuna markets also makes it nearly impossible for local fishermen to strike for higher prices, even when low prices cause serious financial problems for them.

Market Concentration

The vulnerability of regional tuna exporters is also increased by the buyer's domination of the global market. A few buyers in the United States, Japan, and Europe handle so much tuna from so many regions that they have the opportunity and the incentive to accumulate enormous amounts of valuable information. Despite tremendous improvements in market information available to regional producers through organizations like INFOFISH and the Forum Fisheries Agency, the major buyers have a distinct advantage over most regional suppliers in any kind of negotiation.

This domination by buyers also subjects small suppliers to incidents of market misinformation common in any industry where a limited number of "authoritative" sources provide information to everyone else. Because these sources are in a position to influence general sentiment about price trends and supply and demand, they can exploit that position to their own advantage. Alternative sources of information, such as independent consultants and academics, also sometimes have difficulty acquiring or verifying information from nontraditional sources about market changes. Since even "objective" industry and market analysts frequently rely on the few familiar "authoritative" sources of information, they can be—and have been—used as vehicles for disseminating misinformation to suppliers. Major market problems, of course, have their origins in real changes in supply, demand, and price, but contrived market movements are possible, and regional suppliers operating in an oligopsonistic environment (one where few buyers exert disproportionate influence) need to be cautious about the source of their market information.

Price Interdependence

Tuna prices at the ex-vessel level (the price paid to fishermen) and at the wholesale level (the price paid to processors) are usually determined by competitive conditions and price trends at the retail level. In the U.S. market, for example, tuna competes not only with other seafood products but also with poultry and meat products in a highly competitive retail food market. Retailers face their own competitive pressures; they cannot afford to allot shelf space to products that sell slowly or do not allow a reasonable markup. When declining chicken or meat prices or such events as the tuna-porpoise controversy or the fear of mercury or botulism contamination lower the demand for tuna, a reduction in retail prices may be essential to stimulate purchases and protect tuna's share of the food market. Since tuna demand is relatively inelastic with respect to price changes, these price cuts must be dramatic to have an effect. In fact, to increase tuna sales, wholesale price cuts must be drastic enough to allow both a sizable reduction in consumer prices to stimulate sales volume and a sizable retail markup to satisfy retailers and assure that tuna will get adequate promotion and shelf space.

Since the profitability of tuna processing is based on a small margin of profit and a high volume of sales, slashing wholesale prices creates financial problems for processors, who then pass some of these problems back to tuna harvesters in the form of lower ex-vessel tuna prices. In a matter of months, or even weeks, a shift in consumer preference away from canned tuna can be reflected in the prices of canned tuna and raw or frozen tuna on the world market. Again, it is primarily shifting supply conditions that influence tuna prices on the world market, but when demand shifts do occur in retail markets, they are passed on very quickly to wholesale and ex-vessel markets.

Figure 9 shows the interdependence of prices at different market levels. Note that the annual percentage changes in prices in U.S. retail, wholesale, and ex-vessel tuna markets have been remarkably consistent over the past 25 years. Market forces linking prices at these levels originate at the retail level and are felt almost immediately at the wholesale and ex-vessel levels. Access fees paid to Pacific island nations will soon constitute another market level. If these fees are determined by market forces rather than geopolitical maneuvering, they will be influenced by the same market forces as the prices shown in Figure 9 and, with a slight lag, will follow a similar pattern.

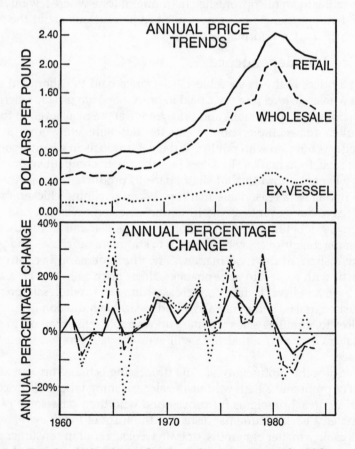

Figure 9. Retail, wholesale, and ex-vessel tuna prices in the U.S. market

CONCLUSION

Tuna from the Pacific island region competes in global markets with tuna from all other ocean areas. Each island nation needs to decide whether it is in its interest to become involved in this global industry and at which market level it should operate, from the selling of access fees to the marketing of raw or frozen tuna, cooked tuna loins or discs, or processed tuna products. Whatever its decision, each nation must be aware of the forces that influence the behavior of competitors and partners. Each nation must also recognize how its tuna-related industries or the exploitation of its tuna resources by foreign operators fits into a tuna production system that includes harvesting, processing, distributing, and marketing operations. Entering a tuna production system at only one stage involves major risks because of the competitiveness of the world tuna industry; hence each island nation must also try to determine where value is added within the production system and at which stage the profits are paid out.

The level of investment required, the risks, and the potential economic payoffs from entering the global tuna market differ at each level. However, unless an island nation can develop a comparative advantage at one or more stages of the production process (harvesting, loining, canning, and so on), it cannot expect to benefit from participating directly in the tuna industry, nor will it be able to attract sincere foreign investors or joint-venture partners. An island nation's proximity to productive fishing areas does not alone provide it with a comparative advantage because other necessities are in short supply in most Pacific island regions.

Tuna harvesting and processing are expensive and risky operations. Because most of them are capital-intensive rather than labor-intensive, they are not likely to generate the kinds of secondary economic activities that can provide the basis for economic development in a remote island setting. Moreover, because most of the materials needed to operate in the tuna industry would have to be imported and because profit margins, even for efficient operations, are small, entering the tuna industry would not generate significant export revenues for most island nations. Lacking a comparative advantage in the harvesting and processing of tuna, most island nations can incur the least risk and derive the greatest benefit by sharing in the profits earned by nations that do have a comparative advantage in the industry. This end is best achieved by collecting access fees or some other form of rent from foreign tuna producers and using the proceeds to develop local industries consistent with national goals.

Controlling access to tuna fisheries on a regional rather than a national basis has advantages in cost savings and in effectiveness; it also increases the amount that island nations can earn collectively. In time, it may also offer them enough control over supplies to have some impact on global markets and on tuna prices. With information about the global tuna trade and a commitment to regional cooperation, island nations could exert considerable power—power that could initiate a shift in control from the buyer's side of the market to the supplier's side.

REFERENCES

Food and Agriculture Organization (FAO)

annual   United Nations Food and Agriculture Organization yearbook of fishery statistics: Catches and landings. Various years. Rome.

annual   United Nations Food and Agriculture Organization yearbook of fishery statistics: Fishery commodities. Various years. Rome.

annual   FISHDAB computerized database (GLOBEFISH). Various years. Rome.

FOODNEWS

weekly   Newsletter. Various years. London.

INFOFISH.

annual   Trade news and marketing digest. Kuala Lumpur.

Selling Areas Marketing, Inc.

annual   Statistics from secondary sources. Various years. Santa Monica, California.

U.S. National Marine Fisheries Service.

annual   Fisheries of the United States. Various years. Washington, D.C.

annual   Market news reports. Various years. Terminal Island, California.

annual   Operation pricewatch. Various years. Washington, D.C.

PART V. FUTURE DIRECTIONS

17.
Prospects and Directions in the Tuna Fishery

David J. Doulman

INTRODUCTION

In Fiji in 1984, at the opening session of tuna treaty negotiations with the United States, the speaker for the Pacific islands delegations said, "Fish are our lives." His simple statement encapsulates the place of fish in the lives of Pacific island peoples: fish are the essence of their subsistence and potentially of their commerce.

Of the many species of fish in the region, tuna is the most abundant, the most valuable, and the most important in world commerce. The fact that about 60 percent of the world's tuna for canning is taken in the region—along with large quantities of non-canning species—shows the scale of the region's fishery in the international tuna trade.

The fleets of Pacific island countries and of distant-water fishing nations (DWFNs) exploit the region's tuna resources. Island governments use license revenue from DWFNs to support expenditure programs; domestic industries contribute substantially to the economies of some countries. DWFNs rely on the robustness of the fishery to deploy their fleets productively.

Past relations between Pacific island countries and DWFNs have been mixed. Although U.S. policy rejected island countries' claims to the tuna resources within their exclusive economic zones (EEZs), some U.S. fishermen acknowledged those claims and paid fees to fish there. Other DWFNs accepted the legitimate claims of island countries to their resources, but their fishermen were sometimes reluctant to deal openly and pay equitably for the tuna they harvested. Since 1984, however, relations between resource-owning countries and resource-harvesting countries have improved.

The Soviet Union's interest in the region, which culminated in agreements with Kiribati (1985) and Vanuatu (1987), and the U.S. government's wish not to permit industry interests to further prejudice international relations in the region led to U.S. agreement on a tuna treaty with island countries (1986). This treaty acknowledges the jurisdiction of Pacific island countries over their tuna resources; furthermore, the U.S. government undertakes to ensure that U.S. fishermen adhere to the treaty's provisions and to impose sanctions if fishermen violate them. This recognition of resource ownership brings U.S. policy into line with the policies of most other coastal states. The treaty is expected to improve relations between the United States and island countries; it may also strengthen the Soviet Union's determination to push through on its initiatives in the region.

Out of sheer economic necessity, island countries consider access requests from any DWFN as a means of increasing the number of fleets operating in the region. The revenue to be paid under terms of the U.S. tuna treaty will reduce the need for countries to seek access agreements with DWFNs that have not traditionally fished in the region. But the pragmatism of Pacific island governments is such that they will deal with any DWFN if, in the words of Prime Minister Ratu Sir Kamisese Mara of Fiji, "the price is right."

Largely as a result of efforts by the Forum Fisheries Agency (FFA), member nations of the South Pacific Forum (SPF) have been able to accumulate information about the operations of DWFN fleets. This information has empowered island countries to negotiate better access arrangements and has reduced the capacity of DWFNs to conceal information. The consequence is that island countries have achieved better financial returns from the tuna harvested by DWFN fleets. However, island countries maintain that there is scope for improvement, and they will continue to press for a fairer share of the revenues generated by the fishery.

In their efforts to achieve this goal, countries are likely to further regulate and control the exploitation of their tuna resources, to encourage linkages between the distant-water fishery and their economies, and to foster the development of domestic industries.

Most Pacific island countries and territories want to be more directly involved in their region's tuna fishery. Some of them already have economically important tuna industries, and despite difficulties in the international tuna industry since 1981 and the loss of some of their foreign partners, the region's domestic industries have survived. As conditions in the international tuna industry stabilize, island countries are expected to promote vertically integrated industries so as to have a hand in every stage, from harvesting through processing to marketing. The collapse of some industries was due partly to the fact that the countries exerted little control over the partnerships, either by investing in them or by making comprehensive, enforceable agreements.

Critics of industry development argue that to achieve a comparative advantage, island countries should limit their involvement in the fishery to licensing DWFN fleets. But such well-intentioned conventional economic wisdom ignores the political importance of tuna in the region and the desire of countries to reduce their dependence on "handouts." To Pacific island countries, tuna is what oil is to countries of the Middle East. And just as the oil-producing countries in the 1970s sought to control their principal resource to increase their benefits, Pacific island countries seek to control their resource to increase their benefits.

This paper offers some forecasts for developments in the Pacific islands tuna fishery during the next few years. It predicts directions in the region's distant-water operations, its domestic industries, and its cross-national cooperation. Although radical departures from present trends are not foreseen, resource-owning countries will probably institute new measures to serve their goals of economic development and economic security.

DISTANT-WATER OPERATIONS

Distant-water tuna-fishing operations in the Pacific islands will remain at a high level because island countries want cooperative arrangements with DWFNs to harvest their tuna resources, and DWFNs want access to the region's fishery. However, vessel licensing—the simplest form of a resource tax—provides little domestic interaction with island economies and generates few secondary benefits for island countries.

Island countries aim to increase their share of benefits from the exploitation of tuna by DWFN fleets, both directly in revenue from access fee payments and indirectly from increased secondary benefits. The economic impact of secondary benefits—for example, the purchase of provisions or fuel oil at ports in the region—is generally more important than the primary impact of access fee payments. These secondary benefits now accrue to the DWFNs where the fleets are based or to ports where tuna is transshipped.

Access

Pacific island countries have generally maintained an open-door policy with respect to tuna access because (1) they could not fully exploit their tuna resources, (2) they recognized the provisions of Article 56 of the Law of the Sea Convention concerning the utilization of surplus fish stocks, (3) they wanted to raise revenue from the sale of fishing licenses, and (4) they preferred not to discriminate between DWFNs in granting licenses.

In a departure from past policy, SPF member countries are expected to implement measures to limit fisheries access, primarily for financial reasons. Some industry commentators claim that access fees should not be raised because the tuna fishery (particularly the purse seine component) is not now yielding resource rent—that is, an excess to be used for the benefit of the citizens who collectively own the resource. Because too many seiners are operating in the fishery, their operations are financially marginal; hence, say these commentators, island countries should not impose a hardship on fishermen by raising fees. But this fact is of no concern to the resource owners, and it should not weigh heavily in their decisions about access fee levels. When a fishery is not yielding resource rent, economic theory prescribes that the level of effort in the fishery be reduced; only then will rent be generated. The proposed reduction in effort will financially benefit both the resource owners and the fishermen remaining in the fishery.

Initial restrictions may be introduced in the purse seine fishery, later ones in the longline fishery. Competition for licenses (through preferential licensing arrangements) should exert upward pressure on fees. It may also lead to an increase in illegal fishing. But with the support of DWFN governments, the sanctions of the FFA's regional vessel register, and perhaps the introduction of rewards in the licensing system for vessels that report vessels known or suspected to be operating without licenses, problems of surveillance and enforcement could be minimized.

Agreements

Most access agreements in the Pacific islands region are bilateral ones concluded with DWFN industry groups. However, concluding agreements on a government-to-government basis has advantages because diplomatic measures can be used to help ensure compliance. Some Pacific island countries (Papua New Guinea among them) are keen to conclude agreements with DWFN governments. However, some DWFN governments (for example, Japan) have indicated that they do not want to be officially involved. Such unwillingness is seen as an attempt to abdicate responsibility: DWFN governments can interfere with the negotiation and administration of agreements without being legally responsible for their actions.

In the purse seine fishery, regional agreements are now an established principle. After the administration of the U.S. tuna treaty is reviewed, the regional licensing concept may be extended to the operations of other purse seine fleets in the region. A single multilateral government-to-government treaty could be concluded for all DWFNs and for all island countries with a stake in the purse seine fishery—

most likely an agreement by the governments of Korea, Japan, the Philippines, and possibly the Soviet Union with the Nauru Group. The Taiwanese fleet could be covered by a parallel industry agreement. Such an approach to licensing DWFN fleets is consistent with the FFA's harmonization objectives, and particularly with the Nauru Group's goals.

In addition to the standard provisions in current access agreements, provisions relating to tuna transshipment could be considered for incorporation in new agreements.

Fleets

The DWFN longline fleet is the largest fleet operating in the Pacific islands region; it is also the most geographically diverse. Longline vessels—primarily from Japan, Korea, and Taiwan—operate in most EEZs; there is also a concentration in the EEZs of the Micronesian countries, Papua New Guinea, and Solomon Islands.

Most of the longliners fishing in the equatorial areas of the central and western Pacific are Japanese vessels fishing primarily for the sashimi market. The Korean and Taiwanese fleets tend to fish in the more temperate areas of the Pacific islands, targeting on albacore (*Thunnus alalunga*) for canning. However, more and more of the vessels are fishing for sashimi-grade tuna. For financial reasons, Japanese corporations are facilitating the change in some cases. The Korean and Taiwanese product enjoys a high degree of market acceptability in Japan, usually fetching as good a price as the Japanese product does.

The longline fishery is the most valuable fishery in the Pacific islands, though it ranks second to the purse seine fishery in volume. Because DWFN access fees are linked to the estimated value of catches, the longline fishery accounts for the largest share of fees paid to island countries. Its dominance is expected to continue.

No major expansion is anticipated in longline fishing in the islands region. Japan's sashimi market is relatively stagnant, and there is increasing competition from foreign producers to supply it. In order to restrict supply and support prices, Japan tightly regulates its longline industry and discourages overseas transshipment, which would increase the supply by reducing the time vessels spend in transit.

The DWFN pole-and-line fleet operating in the islands region consists only of Japanese vessels. The size of the fleet declined during the 1970s and 1980s and is likely to decline further, for technical and financial reasons. The fleet's high operating costs, the problems of acquiring and maintaining live baitfish on fishing trips, and the government's program for replacing pole-and-line vessels with purse seiners have contributed to the demise of this fleet. The distant-water pole-and-line fleet operating in the Pacific islands region is estimated at under

100 vessels. Within the next five years the fleet could be so reduced as to become unimportant. As vessels reach the end of their economic life, they are unlikely to be replaced by new pole-and-line vessels.

The purse seine fleet operating in the region will continue to target the EEZs of countries where they can fish throughout the year. Further internationalization of the fishery is expected, and competition for resource access between individual vessels and between fleets of different registrations may develop as a result of restrictions imposed by island countries.

The Japanese fleet will probably remain at its current size, but its capacity will be increased. This trend is already established: as seiners are fully depreciated, they will be replaced by larger vessels. Within the next five years the entire Japanese purse seine fleet will be replaced. As increased capacity allows vessels to remain on the fishing grounds for longer periods, unproductive traveling time will be reduced and total catches will rise.

Korean and Taiwanese purse seine fleets are expanding. The trend has been to acquire or construct superseiner-class vessels, but more recently the Taiwanese have opted to build vessels of the same size and specification as the new Japanese vessels. Although the Korean government previously regulated purse seine licenses, it now appears that controls have been lifted. The Korean and Taiwanese fleets, like the Japanese fleet, will make the Pacific islands region their principal area of operation.

The U.S. purse seine fleet will continue to regard the Pacific islands as secondary to the eastern tropical Pacific, and the Pacific islands region will decline in relative importance for the U.S. fleet because of reductions in its size. No new U.S. investment in purse seine vessels is envisaged in the medium term because the fleet became heavily overcapitalized in the 1970s and many vessel owners are now at or near bankruptcy.

However, the number of U.S.-owned seiners operating in the region under flags of convenience is expected to rise. If regional restrictions are imposed on the purse seine fleet, vessels from these flag states could be excluded first because their governments (for example, Panama and Cayman Islands) are unlikely to enter into appropriate access agreements with Pacific island countries.

Transshipment

The Law of the Sea Convention (Article 62) permits a coastal state to require DWFN fleets to offload or to transship their catches made in its EEZ at domestic ports. The intent of this provision is to generate economic benefits for the resource owners from DWFN fleet activities. An estimated 100,000 tonnes of purse-seine-caught tuna is being trans-shipped in the Pacific islands annually. However, transshipment is not benefiting the countries in whose EEZs the fish is harvested.

Tuna transshipment in the Pacific islands currently does not use shore-based infrastructure; it is confined to vessel-to-vessel transfers. Several island countries—notably the Federated States of Micronesia and Papua New Guinea—have expressed interest in establishing trans-shipment facilities. These facilities could involve vessel-to-vessel trans-fers at first; shore-based facilities could be built later if financially feasible.

Purse seine vessels may in future be induced or required to trans-ship some catches in licensing countries as a condition of license. For example, preferential access might be given to DWFN vessels prepared to transship at domestic ports, and vessel owners could be assured of license renewal immediately following transshipment. They could then spend longer periods fishing, and Pacific island countries would benefit from the increase in commercial port activity.

The need for introducing transshipment provisions in access agree-ments is highlighted by recent industry reports concerning purse seine transshipment at sea. Previously it was thought that catches could be transshipped at sea only under very favorable weather conditions, but it now seems that the technique has been perfected for use even in rough weather. But transshipment at sea undermines the principles upon which access agreements and trip license fees are predicated, and it may be necessary for Pacific island countries to introduce a licensing condition requiring seiners to transship their catches at domestic ports in order to prevent this practice.

Overseas tuna transshipment will be resisted by Japan. The govern-ment's policy requires tuna landed in the Pacific islands to be discharged by the harvesting vessel at a Japanese port. However, some sections of Japanese industry are believed to be at odds with their government on this issue. But in this and other issues relating to DWFN fleet oper-ations in the Pacific islands, the interests of the resource-owning coun-tries, not those of the DWFNs, must take precedence.

Aid

Some DWFN governments have conventionally offered aid to island countries to achieve more favorable terms and conditions of access. However, members of the SPF have agreed that aid and access should be delinked. It is now generally accepted that, while fisheries aid is welcome, access to the tuna fishery must be based on commercial con-siderations alone.

Aid provided by Japan has benefited some island countries in developing fishing fleets, training fishermen, and providing fishing gear. But there have also been instances where inappropriate and inadequate assistance was provided, primarily when the assistance sought was of a commercial nature with potential implications for Japanese industry.

The purpose of aid is to help island countries improve their artisanal and commercial fisheries capabilities. Assistance that is technically outdated or unsuitable or aid that frustrates fisheries progress is shortsighted and regarded by some observers as morally wrong.

A fisheries aid program recently instituted by the United States has proven to be of high quality. Island countries have been closely consulted about their development and aid priorities, and the United States has genuinely attempted to meet those priorities. Aid to be provided under the U.S. tuna treaty is expected to be of similar quality.

As part of their efforts to replace foreign fisheries personnel, Pacific island countries attach high priority to fisheries training for their nationals. There are several institutions in the region for this purpose, funded by island governments, sometimes with assistance from foreign governments. However, as a means of boosting the resources available to these training institutions, an "education levy" might be applied to each tonne of tuna harvested by DWFN fleets. Some African countries impose such a levy, earmarking it for fisheries training and education. Thus an international precedent for the levy exists, and Pacific island countries could reasonably require DWFNs to contribute to fisheries training and education in proportion to the quantity of tuna harvested by their fleets.

DOMESTIC INDUSTRIES

Several Pacific island countries and territories already have well-established tuna industries. In the medium term it is expected that these industries will be strengthened and that tuna processing will be expanded.

Fleet development

Longline, pole-and-line, and purse seine fleets are based in the Pacific islands region. In terms of vessel numbers, employment, and economic benefits, the pole-and-line fleet is the most important.

Approximately 40 pole-and-line vessels are currently based in five SPF countries, down from about 80 pole-and-line vessels in the early 1980s. The 1982 collapse of the domestic fishery in Papua New Guinea was responsible for the decline in vessel numbers. Most of the region's pole-and-line fleet—about 75 percent—now operates in Solomon Islands.

Expansion of pole-and-line fishing is constrained by the availability of suitable baitfish. For this reason not all island countries are equally placed to engage in this type of fishing. Moreover, pole-and-line production costs are generally higher than purse seine production costs; because the two methods are competing for the same end use—canning—the odds are financially against pole-and-line operations. However, pole-and-line fishing does have advantages: it uses an intermediate level of technology and, because it is the most labor-intensive method of tuna fishing, it maximizes employment benefits for Pacific islanders.

No significant expansion of pole-and-line fishing is expected in the island countries that are expanding or developing skipjack tuna (*Katsuwonus pelamis*) fisheries; they will most certainly opt to deploy purse seine vessels.

Only two island countries—Nauru and Solomon Islands—have invested in purse seine fleets, although three Ecuadorian-owned vessels are registered in Vanuatu and about 30 foreign seiners are based more or less permanently in the region, mainly in American Samoa and Guam. Little is known about the Nauruan seiners, but it is believed they are not fishing. The Solomon Islands vessels, each of 499 gross registered tonnes and of Japanese design, are being built in Australia for delivery in late 1987.

Several island countries are considering ventures requiring investment in purse seiners, but enthusiasm for fleet investment has been dampened by the instability in the international tuna industry since 1981, and they are proceeding cautiously. Nonetheless, several countries will probably proceed with the development of purse seine fleets of smaller vessels that can be accommodated by existing infrastructure. They will probably operate in a similar way to the seiner *Dae Bong 2*, which fished successfully from Rabaul in Papua New Guinea. The smaller seiners will not be extremely mobile high-seas vessels; they will fish inshore and closer to transshipment points (for example, in the Bismarck and Solomon seas) than do the DWFN fleets.

Domestic longline fleets have operated in the Pacific islands since the mid-1970s, and in recent years they have been based in three countries and territories—Solomon Islands, New Caledonia, and Tonga. In addition, foreign longline fleets from Japan, Korea, and Taiwan have been based in the Pacific islands (American Samoa, Fiji, French Polynesia, and Vanuatu) since the 1950s. These fleets, usually contracted to canneries in American Samoa, Hawaii, and Fiji, fish for albacore.

Scope for Pacific island countries to participate in the longline fishery is available but limited. Micronesian countries could deploy sashimi longliners and use transshipment vessels. According to Japanese industry reports, sashimi-grade tuna that is well handled and transshipped outside Japan is competitive with fish landed directly at Japanese ports. An astutely managed fleet producing good sashimi should be able to operate profitably.

In the albacore longline fishery, prospects for fleet investment are uncertain. However, U.S. government-supported surveys in 1986 in the southern albacore fishery using trolling vessels produced encouraging results. A survey report concluded that it is economically feasible for U.S. trollers to operate seasonally in the fishery and that the albacore population there is in good condition and able to support a surface fishery.

Pacific island countries wanting to develop an albacore fleet might investigate trolling rather than longlining. Trollers could be deployed from American Samoa, the Cook Islands, French Polynesia, Tonga, and Western Samoa, and fish could be landed at Pago Pago for processing.

Processing

Tuna processing can involve the production of *katsuobushi*, tuna loins, and canned product. Currently there is no tuna loining in the Pacific islands, but *katsuobushi* production and tuna canning are going on at several locations.

As a matter of policy, Pacific island countries are committed to expanding existing tuna-processing industries and promoting new ones. Countries are pursuing this goal as a means of broadening their industrial and economic bases and gaining increased benefits from their tuna fisheries.

American Samoa's two canneries have been recently upgraded and expanded, and plans are being made to relocate Solomon Islands' cannery and at least double its processing capacity. Fiji also plans to upgrade its cannery and, if possible, to increase its output. The Federated States of Micronesia and Papua New Guinea are also considering tuna processing.

Island countries are becoming more self-confident of their capacity to invest in fishing and processing ventures. Although multinational tuna companies dominate the international scene, they do not have a monopoly on market and industry information. Small, vertically integrated ventures can succeed so long as they deliver a high-quality product and select their markets to avoid competition with the large producers. These producers—especially Thai and U.S. canners—primarily target the mass U.S. market. In comparison to other markets (for example, some European markets) and some segments of the U.S. market, the mass U.S. market is price-sensitive, not quality-sensitive. Therefore, as a general marketing strategy, it would be imprudent for the non-U.S. canneries in the Pacific islands to try to compete on the U.S. tuna market with Thai, Filipino, and U.S. canned products because canned tuna from Fiji and Solomon Islands, for example, is much superior in quality. Smaller countries are able to trade on quality in discriminating markets, and they should focus on those markets.

Of all Pacific island countries, Papua New Guinea probably has the best potential to develop a tuna-canning industry because it has the population, natural endowments, availability of fish, ability to service fleets, and capacity to implement industrial-scale projects. With a tuna transshipment and processing facility, Papua New Guinea could have a tuna industry that rivals in economic importance some of its existing agricultural industries.

The possibility of establishing a regional tuna cannery has merit for smaller Pacific island countries with restricted opportunities for industrial development. This possibility is expected to be evaluated by some SPF member countries.

Prospects for the expansion of *katsuobushi* processing and the production of tuna loins in the region are not bright for the immediate future. *Katsuobushi* production is limited by the size of Japan's market and the lack of firewood in the region for smoking the product.

Logistical problems and technical difficulties associated with the production of tuna loins in the region need to be addressed and solved if this type of processing is to be pursued. It holds potential advantages for island countries and canneries located in high-wage countries. Attempts to produce skipjack loins on a trial basis in Solomon Islands for the Japanese market are reported to have been technically but not financially feasible. However, some industry commentators contend that loin production could be technically and financially feasible if sufficient attention is devoted to research and development. Further investigation is justified.

REGIONAL COOPERATION

Regional cooperation on fisheries is firmly established among SPF countries. Since 1980 they have achieved considerable gains by cooperating. SPF member countries will continue to meet regularly and to consult and exchange information, to plan common courses of action, and to harmonize terms and conditions of access for DWFN fleets. Further harmonization of terms and conditions will be of special concern to Nauru Group members. Members are likely to conclude additional implementing arrangements as the need arises.

Pacific island countries are acutely aware that individually they have limited strength in dealing with DWFNs but that collectively they can accomplish their ends. In pursuing cooperative arrangements, they also acknowledge that it is not in their self-interest to jeopardize relations by making radical and unreasonable demands on DWFNs. If relations are jeopardized, island countries stand to lose an important source of revenue.

It is expected that DWFNs and Pacific island countries will seek closer cooperation. All DWFNs now recognize the jurisdiction of island countries over their tuna resources. With the support of DWFN governments, incidents of illegal fishing in the region will no doubt decline.

Some U.S. officials believe that U.S. domestic legislation (Fisheries Conservation and Management Act of 1976) will in time be amended to bring U.S. tuna policy formally into line with internationally accepted policy. Such a change would be welcomed by U.S. Pacific territories, Hawaii, and U.S. east coast fishermen. The change could also help the United States improve its relations with Latin American countries.

The FFA was established in 1979 primarily to assist SPF countries in implementing changes resulting from the Law of the Sea negotiations and to help members make the most of the new opportunities that extended jurisdiction provided. Initially, the FFA enabled island countries to present a united, consistent, and informed front in negotiations with DWFNs. The initial objective for FFA's establishment has been largely achieved. The question now is, what role will the FFA play in the future?

Clearly, the FFA will have to help island countries further coordinate and harmonize DWFN access arrangements. However, because not all island countries have a real and equal interest in DWFN activity, the FFA is likely to become more closely involved in promoting domestic industries. The agency could play a leading role in the investigation and possible establishment of a regional processing facility.

The islands have the basis for developing a regional industry. Fleets from several countries deliver catches to Fiji's cannery, and such cooperation could be extended. A regional tuna cannery makes sense, but the selection of a site is likely to be sensitive. Not all island countries are well placed for tuna processing. And although political considerations are important, they should be balanced against the need to locate a commercial venture where its prospects for financial success are greatest.

In addition to existing arrangements in the Pacific islands for fisheries collaboration (the FFA for SPF member countries and the Western Pacific Regional Fisheries Management Council for U.S. Pacific territories), there is merit in broader informal consultation between island countries and territories. Such consultations could improve the flow of information between countries and territories that have commercial ties but no formal association. Tuna harvested in the EEZs of the Federated States of Micronesia and Papua New Guinea is transshipped at Guam and Tinian; tuna harvested in Solomon Islands is processed in American Samoa. It could benefit such countries as the Federated States of Micronesia, Papua New Guinea, and Solomon Islands to confer on tuna matters with officials from American Samoa, Guam, and the Northern Marianas.

It would be inappropriate for the FFA to organize such a broadly based consultation. However, it could be sponsored by the South Pacific Commission, perhaps following its annual fisheries technical meeting. If appropriate, government officials and industry representatives from DWFNs could also be invited to attend all or part of the consultation.

CONCLUSION

Pacific island countries and territories propose to increase the benefits they receive from the exploitation of their tuna resources by increasing their financial returns from licensing DWFN fleets and by promoting the expansion and development of domestic industries. Island countries perceive these increases as necessary for obtaining an equitable share of the benefits flowing from the tuna fishery and for bolstering their dependent and sluggish economies.

Following the declaration of extended jurisdiction, most island countries adopted an open-door policy on licensing DWFN fleets. However, this policy is likely to be changed gradually. Some fleets may be required to start transshipping catches at domestic ports as a means of strengthening linkages between distant-water fisheries and island economies.

Despite these possible changes, DWFN operations will remain at a high level, simply because they benefit both island countries and DWFNs. Both sides gain from stable access arrangements.

The changes in the international tuna industry between 1981 and 1986 have been particularly instructive for Pacific island countries. The countries that have lost industries have learned that to ensure industry survival, they must be more active in controlling them, because the motives of foreign partners (acquisition of raw material) and those of island governments (stable and viable industry development) often do not coincide. This problem is not confined to the fishing industry; it is symptomatic of all extractive-resource-based industries.

Prospects for the tuna industry worldwide are uncertain. Island countries are aware of the risks of investing in fleets and in tuna processing; hence they will be cautious in investing.

Through the FFA and the Western Pacific Regional Fisheries Management Council, island countries and territories will continue to cooperate on fisheries matters. Having already achieved a high degree of harmonization in their relations with DWFNs, SPF countries are likely to enter a period in which the FFA will focus on assisting in developing domestic industry.

Because Pacific island countries are determined to be more than industry spectators, they will regulate the region's tuna fishery. Cooperation between DWFNs and island countries will be enhanced if DWFNs accept that the tuna fishery is more than just a source of raw material for industries outside the region to process. By cooperating with island countries and helping them to achieve their development goals, DWFNs will also serve their own long-term interests.

Contributors

PARZIVAL COPES, professor of economics and director, Institute of Fisheries Analysis, Simon Fraser University, Vancouver, B.C., Canada.

DAVID J. DOULMAN, project director, Multinational Corporations in the Pacific Tuna Industry, Pacific Islands Development Program, East-West Center, Honolulu, HI, USA.

AUGUST FELANDO, president, American Tunaboat Association, San Diego, CA, USA.

NORIO FUJINAMI, special adviser on international affairs to the minister of Agriculture, Forestry, and Fisheries, Tokyo, Japan.

ALFONSO P. GALEA'I, director, Office of Development Planning and Tourism, Pago Pago, American Samoa.

ROBERT GILLETT, fisheries development advisor, FAO/UNDP South Pacific Regional Fisheries Development Program, Suva, Fiji.

FLORIAN GUBON, legal officer, International Law Branch, Department of Justice, Waigani, Papua New Guinea.

LINDA LUCAS HUDGINS, professor of economics, University of Notre Dame, Notre Dame, IN, and fellow at Pacific Islands Development Program, East-West Center, Honolulu, HI, USA.

ANTHONY V. HUGHES, governor, Central Bank of Solomon Islands, Honiara, Solomon Islands.

DENNIS M. KING, research director, Economic Research Group Pacific, Inc., San Diego, CA, USA.

YOSHIAKI MATSUDA, professor, Faculty of Fisheries, Kagoshima University, Kagoshima, Japan.

CAROLYN NICOL, student, William S. Richardson School of Law, University of Hawaii at Manoa, Honolulu, HI, USA.

SAMUEL G. POOLEY, industry economist and acting leader, Fisheries Management Research Program, National Marine Fisheries Service (Southwest Fisheries Center), Honolulu, HI, USA.

MICHAEL J. RIEPEN, fisheries consultant, Raumati Beach, New Zealand.

DONALD M. SCHUG, planner, Office of Development Planning and Tourism, Pago Pago, American Samoa.

JOHN SIBERT, coordinator, Tuna Program, South Pacific Commission, Noumea, New Caledonia.

ANTHONY J. SLATYER, legal officer, Department of Primary Industry, Canberra, ACT, Australia.

FOUA TOLOA, director, Office of Tokelau Affairs, Apia, Western Samoa.

JON VAN DYKE, professor, William S. Richardson School of Law, University of Hawaii at Manoa; adjunct research associate, Resource Systems Institute, East-West Center; member, executive board of the Law of the Sea Institute, Honolulu, HI, USA.

NOTES

NOTES